Basics of
BIOSTATICS

Basics of
BIOSTATICS

Rohit Garg

ANMOL PUBLICATIONS PVT. LTD.
NEW DELHI-110 002 (INDIA)

ANMOL PUBLICATIONS PVT. LTD.
Regd. Office: 4360/4, Ansari Road, Daryaganj,
New Delhi-110002 (India)
Tel.: 23278000, 23261597, 23286875, 23255577
Fax: 91-11-23280289
Email: anmolpub@gmail.com
Visit us at: www.anmolpublications.com

Branch Office: No. 1015, Ist Main Road, BSK IIIrd Stage
IIIrd Phase, IIIrd Block, Bengaluru-560 085 (India)
Tel.: 080-41723429 • Fax: 080-26723604
Email: anmolpublicationsbangalore@gmail.com

Basics of Biostatics

First Edition, 2011

ISBN 978-81-261-4906-3

PRINTED IN INDIA

Printed at Suman Printers, Delhi

Contents

Preface *vii*

1. Logarithms 1
2. The Arithmetic Mean 30
3. Frequency Discordance 47
4. Variation Measurement 66
5. Measurement Scale 84
6. Moments and Distribution Characteristics 94
7. Measures of Central Tendency 113
8. Measures of Variability 129
9. Analysis of Variance 147
10. The Histogram 183
11. The Binomial Distribution 202
12. Sampling Methods 219
13. Measure and Probability Theory 253

Bibliography 293
Index 295

Preface

Biostatistics is the application of statistics to a wide range of topics in biology. The science of biostatistics encompasses the design of biological experiments, especially in medicine and agriculture; the collection, summarization, and analysis of data from those experiments; and the interpretation of, and inference from, the results.

It is the science of statistics applied to the analysis of biological or medical data. Biostatistics in the public health context consists primarily of developing descriptive statistics describing the overall health and well being of a population. These statistics include such measures as birth, death, and infant death rates; disease incidence and prevalence; and trends of this data over time.

Biostatistics can be defined as the application of the mathematical tools used in statistics to the fields of biological sciences and medicine. Biostatistics is a growing field with applications in many areas of biology including epidemiology, medical sciences, health sciences, educational research and environmental sciences.

Author

1

Logarithms

Logarithms represent a specialized use of exponents. By means of logarithms, computation with large masses of data can be greatly simplified. For example, when logarithms are used, the process of multiplication is replaced by simple addition and division is replaced by subtraction. Raising to a power by means of logarithms is done in a single multiplication, and extracting a root reduces to simple division.

DEFINITIONS

In the expression $2^3 = 8$, the number 2 is the base (not to be confused with the base of the number system), and 3 is the exponent which must be used with the base to produce the number 8. The exponent 3 is the logarithm of 8 when the base is 2. This relationship is usually stated as follows: The logarithm of 8 to the base 2 is 3. In general, the logarithm of a number N with respect to a given base is the exponent which must be used with the base to produce N. Table-1 illustrates this.

Table: Logarithms with Various Bases

$2^3 = 8$	$\log_2 8 = 3$
$4^2 = 16$	$\log_4 16 = 2$
$5^0 = 1$	$\log_5 1 = 0$
$27^{2/3} = 9$	$\log_{27} 9 = 2/3$

Table shows that the logarithmic relationship may be expressed equally well in either of two forma; these are the exponential form and the logarithmic form. Observe, that the base of a logarithmic expression is indicated by placing a subscript just below and to the right of the abbreviation "log." Observe also that the word "logarithm" is abbreviated without using a period.

The equivalency of the logarithmic and exponential forms may be used to restate the fundamental definition of logarithms in its most useful form, as follows:

$$b^X = N \text{ implies that } \log_b N = x$$

In words, this definition is stated as follows: If the base b raised to the x power equals N, then x is the logarithm of the number N to the base b. One of the many uses of logarithms may be shown by an example in which the base is 2. Table 8-2 shows the powers of 2 from 0 through 20. Suppose that we wish to use logarithms to multiply the numbers 512 and 256, as follows:

From table,

$$8\text{-}2,\ 512 = 2^{9,}\ 256 = 2^8$$

Then $512 \times 256 = 2^9 \times 2^8 = 2^{17}$ and from the table again 2^{17} = 131072. It is seen that the problem of multiplication is reduced to the simple addition of the exponents 9 and 8 and finding the corresponding power in the table. Table shows the base 2 in the exponential form with its corresponding powers. The actual computation in logarithmic work does not require that we record the exponential form. All that is required is that we add the appropriate exponents and have available a table in which we can look up the number corresponding to the new exponent after adding. Therefore, table is adequate for our purpose.

Solving the foregoing example by this table, we have the following:

$\log_2 512 = 9$

$\log_2 256 = 8$

$\log_2$ of the product = 17

Therefore, the number we seek is the one in the table whose logarithm is 17. This number is 131,072. In this example, we found the exponents directly, added them since this was a Multiplication problem, and located the correspending power. This avoided the unnecessary step of writing the base 2 each time.

NATURAL AND COMMON LOGARITHMS

Many natural phenomena, such as rates of growth and decay, are most easily described in terms of logarithmic or exponential formulas. Furthermore, the geometric patterns in which certain seeds grow (for example, sunflower seeds) is a logarithmic spiral.

These facts explain the name "natural logarithms." Natural logarithms use the base e, which is an irrational number approximately equal to 2.71828. This system is sometimes called the Napierian system of logarithms, in honour of John Napier, who is credited with the invention of logarithms. To distinguish natural logarithms from other logarithmic systems the abbreviation, In, is sometimes used. When In appears, the base is understood to be e and need not be shown. For example, either log, 45 or In 45 signifies the natural logarithm of 45.

COMMON LOGARITHMS

As has been shown in preceding paragraphs, any number may be used as a base for a system of logarithms. The selection of a base is a matter of convenience. Briggs in 1617 found that base 10 possessed many advantages not obtainable in ordinary calculations with other bases. The selection of 10 as a base proved so satisfactory that today it is used almost exclusively for ordinary calculations. Logarithms with 10 as a base are therefore called Common Logarithms. When 10 is used as a base, it is not necessary to indicate it in writing logarithms.

For example,

$$\log 100 = 2$$

is understood to mean the same as,

$$\log_{10} 100 = 2$$

If the base is other than 10, it must be specified by the use of a subscript to the right and below the abbreviation "log." As noted in the foregoing discussion of natural logarithms, the use of the distinctive abbreviation "In" eliminates the need for a subscript when the base is *e*. It is relatively easy to convert common logarithms to natural logarithms or vice versa, if necessary. It should be noted further that each system has its peculiar advantages, but for most everyday work, the common system is more often used. A simple relation connects the two systems. If the common logarithm of a number can be found, multiplying by 2.3026 gives the natural logarithm of the number.

For example,

$$\log 1.60 = 0.2041$$

$$\ln 1.60 = 2.3026 \times 0.2041 = 0.4700$$

Thus the natural logarithm of 1.60 is 0.4700, correct to four significant digits. Conversely, multiplying the natural logarithm by 0.4343 gives the common logarithm of a number. As might be expected, the conversion factor 0.4343 is the reciprocal of 2.3026.

This is shown as follows:

$$\frac{1}{2.30626} = 0.4343$$

Positive Integral Logarithms

The derivation of positive whole logarithms is readily apparent. For example, we see in table that the logarithm of 10 is 1. The number 1 is simply the exponent of the base 10 which yields 10. This is shown in table opposite the logarithmic equation.

Similarly,

$10^0 = 1$........... $\log 1 = 0$

$10^2 = 100$......... $\log 100 = 2$

$10^3 = 1{,}000$...... $\log 1{,}000 = 3$

$10^4 = 10{,}000$..... $\log 10{,}000 = 4$

Table. Exponential and Corresponding Logarithmic Notations Using Base 10

A	B
$10^{-4} = \frac{1}{10^4} = 0.0001$	$\log 0.0001 = -4$
$10^{-3} = \frac{1}{10^3} = 0.001$	$\log 0.001 = -3$

$10^{-2} = \frac{1}{10^2} = 0.01$	log 0.01 =–2
$10^{-1} = \frac{1}{10^2} = 0.01$	log 0.01 =–2
$10^{-1/2} = \frac{1}{\sqrt{10}} = \frac{\sqrt{10}}{10} = 0.31623$	log 0.31623 =–0.5 = 0.5–1
$10^9 = 1$	log 1 = 0
$10^{1/2} = \sqrt{10} = 3.1623$	log 3.1623 = 0.5
$10^1 = 10$	log 10 = 1
$10^{3/2} = 10\sqrt{10} = 31.623$	log 31.623 = 1.5
$10^2 = 100$	log 100 = 2
$10^{5/2} = 10^2\left(\sqrt{10}\right) = 316.23$	log 316.23 = 2.5
$10^3 = 1{,}000$	log 1,000 = 3
$10^{7/2} = 10^3\left(\sqrt{10}\right) = 3162.3$	log 3162.3 = 3.5
$10^4 = 10{,}000$	log 10,000 = 4

Positive Fractional Logarithms

Notice that the logarithm of 1 is 0 and the logarithm of 10 is 1. Therefore, the logarithm of a number between 1 and 10 is between 0 and 1.

An easy way to verify this is to consider some numbers between 1 and 10 which are powers of 10; the exponent in each case will then be the logarithm we seek. Of course, the only powers of 10 which produce numbers between 1 and 10 are fractional powers.

Example: $10^{1/2} = 3.1623$ (approximately) $10^{0.5} = 3.1623$

Therefore, log 3.1623 = 0.5: Other examples are shown in the table for $10^{3/2}$, $10^{5/2}$, and $10^{7/2}$.

Notice that the number that represents $10^{3/2}$, 31.623, logically enough lies between the numbers representing 10^1 and 10^2 -that is, between 10 and 100. Notice also that $10^{5/2}$ appears between 10^2 and 10^3, and $10^{7/2}$ lies between lo3 and 104.

Negative Logarithms

Table shows that negative powers of 10 may be fitted into the system of logarithms. We recall that 10^{-1} means 1/10, or the decimal fraction, 0.1. What is the logarithm of 0.1?

Solution: $10^{-1} = 0.1$; $\log 0.1 = -1$

Likewise $10^{-2} = 0.01$; $\log 0.01 = -2$

Negative Fractional Logarithms

Notice in table that negative fractional exponents present no new problem in logarithmic notation. For example, 1O'1'2 means,

$$\frac{1}{\sqrt{10}}$$

$$\frac{1}{\sqrt{10}} = \frac{\sqrt{10}}{10} = 0.31623$$

What is the logarithm of 0.316231

Solution: $10^{-1/2} = 0.31623$; $\log 0.31623 = -1/2 = -0.5$

Table shows logarithms for numbers ranging from 0.0001 to 10,000. Notice that there are only 8 integral logarithms in the entire range. Excluding zero logarithms, the logarithms for all other numbers in the range are fractional or contain a fractional part. By the year 1628, logarithms for all integers from 1 to 100,000 had been computed.

Practically all of these logarithms contain a fractional part. It should be remembered that finding the logarithm of a number is nothing more than expressing the number as a power of 10. Table shows the numbers 1 through 10 expressed as powers of 10. Most of the exponents which comprise logarithms are found by methods be- yond the scope of this text. However, it is not necessary to know the process used to obtain logarithms in order to make use of them.

Table: The Numbers 1 through 10 Expressed as Powers of 10

$1 = 10^{0}$	$6 = 10^{0.77815}$
$2 = 10^{0.30103}$	$7 = 10^{0.84510}$
$3 = 10^{0.47712}$	8 = 100.90309
$4 = 10^{0.60206}$	$9 = 10^{0.95424}$
$5 = 10^{0.69697}$	$10 = 10^{1}$

COMPONENTS OF LOGARITHMS

The fractional part of a logarithm is usually written as a decimal. The whole number part of a logarithm and the decimal part have been given separate names because each plays a special part in relation to the number which the logarithm represents. The whole number part of a logarithm is called the Characteristic. This part of the logarithm shows the position of the decimal point in the associated number. The decimal part of a logarithm is called the Mantissa. For a particular sequence of digits making up a number, the mantissa of a common logarithm is always the same regardless of the position of the decimal point in that number. For example, log 5270 = 3.72181; the mantissa is 0.72181 and the characteristic is 3.

CHARACTERISTIC

The characteristic of a common logarithm shows the position of the decimal point in the associated number. The characteristic for a given number may be determined by inspection. It will be remembered that a common logarithm is simply an exponent of the base 10. It is the power of 10 when a number is written in scientific notation. When we write log 360 = 2.55630, we understand this to mean 10 2.55630 = 360. We know that the number is 360 and not 36 or 3,600 be cause the characteristic is 2. We know 10 is 10, 10^2 is 100, and 10^2 is 1,000.

Therefore, the number whose value is $10^{2.55630}$ must lie between 100 and 1,000 and of course any number in that range has 3 digits. Suppose the characteristic had been 1: where would the decimal point in the number be placed? Since 10^1 is 10 and 10^2 is 100, any number whose logarithm is between 1 and 2 must lie between 10 and 100 and will have 2 digits. Notice how the position of the decimal point changes with the value of the characteristic in the following examples:

log 36,000 =4.55630
log 3,600 =3.55630
log 360 =2.55630
log 36 =1.55630
log 3.6 =0.55630

Note that it is only the characteristic that changes when the decimal point is moved. An advantage of using the base 10 is thus revealed: If the characteristic is known, the decimal point may easily be placed. If the number is known, the characteristic may be determined by inspection; that is, by observing the location of the decimal point. Although an understanding of the relation of the characteristic to the powers of 10 is necessary for thorough comprehension of logarithms, the characteristic may be determined mechanically by application of the following rules:

1. For a number greater than 1, the characteristic is positive and is one less than the number of digits to the left of the decimal point in the number.
2. For a positive number less than 1, the characteristic is negative and has an absolute value one more than the number of zeros between the decimal point and the first nonzero digit of the number.

Table contains examples of each type of characteristic.

Table: Positive and Negative Characteristics

Number	Power of 10	Digits in number	Characteristic to the left of decimal point
		Between:	
134	10^2 and 10^3	3	2
13.4	10^1 and 10^2	2	1
1.34	10^0 and 10^1	1	0
			Zeros between decimal point and first non zero digit
0.134	10^{-1} and 10^0	0	–1
0.0134	10^{-2} and 10^{-1}	1	–2
0.00134	10^{-3} and 10^{-2}	2	–3

Negative Characteristics

When a characteristic is negative, such as –2, we do not

carry out the subtraction, since this would involve a negative mantissa. There are several ways of indicating a negative characteristic. Mantissas as presented in the table in the appendix are always positive and the sign of the characteristic is indicated separately. For example, where log 0.023 = $\bar{2}.36173$, the bar over the 2 indicates that only the characteristic is negative-that is, the logarithm is –2 + 0.36173. Another way to show the negative characteristic is to place it after the mantissa. In this case we write 0.36173–2. A third method, which is used where possible throughout this chapter, is to add a certain quantity to the characteristic and to subtract the same quantity to the right of the mantissa.

In the case of the example, we may write:

$$\begin{array}{l} \bar{2}.36173 \\ \underline{10 \qquad -10} \\ 8.36173-10 \end{array}$$

In this way the value of the logarithm remains the same but we now have a positive characteristic as well as a positive mantissa.

MANTISSA

The mantissa is the decimal part of a logarithm. Tables of logarithms usually contain only mantissas since the characteristic can be readily determined as explained previously. Table shows the characteristic, mantissa, and logarithm for several positions of the decimal point using the sequence of digits 4, 5, 6. It will be noted that the mantissa remains the same for that particular sequence of digits, regardless of the position of the decimal point.

Table: Effect of Changes in the Location of the Decimal Point

Number	Characteristics	Mantissa	Logarithm
45,600	4	0.6590	4.6590
4.560	3	0.6590	3.6590
456	2	0.6590	2.6590
45.6	1	0.6590	1.6590
4.56	–1	0.6590	0.6590–1
0.0456	–2	0.6590	0.6590–2
0.00456	–3	0.6590	0.6590–3

Appendix I of this training course is a table which includes the logarithms of numbers from 1 to 100. For our present purpose in using this table, we are concerned only with the first and sixth columns.

The first column contains the number and the sixth column contains its logarithm. For example, if it is desired to find the logarithm of 45, we would find the number 45 in the first column, look horizontally across the page to column 6 and read the logarithm, 1.65321. A glance down the logarithm column will reveal that the logarithms increase in value as the numbers increase in value. It must be noted in this particular table that both the mantissa and the characteristic are given for the number in the first column.

This is simply an additional aid, since the characteristic can easily be determined by inspection. Suppose that we wish to use the table to find the logarithm of a number not shown in the "number" column. By recalling that the mantissa does not change when the decimal point moves, we may be able to determine the desired logarithm. For example, the number 450 does not appear in the number column of the table. However, the number 45 has the same mantissa as 450; the only difference between the two logs is in their characteristics. Thus the logarithm of 450 is 2.65321.

THE SLIDE RULE

In 1620, not long after the invention of logarithms, Edmond Gunter showed how logarithmic calculations could be carried out mechanically. This is done by laying off lengths on a rule, representing the logarithms of numbers, and by combining these lengths in various ways. The idea was developed and with the contributions of Mannheim in 1851 the slide rule came into being as we know it today.

The slide rule is a mechanical device by which we can carry out any arithmetic calculation with the exception of addition and subtraction. The most common operations with the slide rule are multiplication, division, finding the square or cube of a number, and finding the square root or cube root of a number. Also trigonometric operations are frequently performed. The advantage of the slide rule is that it can be used with relative ease to solve complicated problems. One limitation is that it will give results with a maximum of only three accurate significant digits. This is sufficient in most calculations, however, since most physical constants are only correct to two or three significant digits. When greater accuracy is required, other methods must be used. A simplified diagram of a slide rule is pictured in figure. The sliding, central part of the rule is called the Slide. The movable glass or plastic runner with a hairline imprinted on it is called the Indicator. There is a C scale printed on the slide, and a D scale exactly the same as the C scale printed on the Body or Stock of the slide rule. The mark that is associated with the primary number 1 on any slide rule scale is called the Index. There is

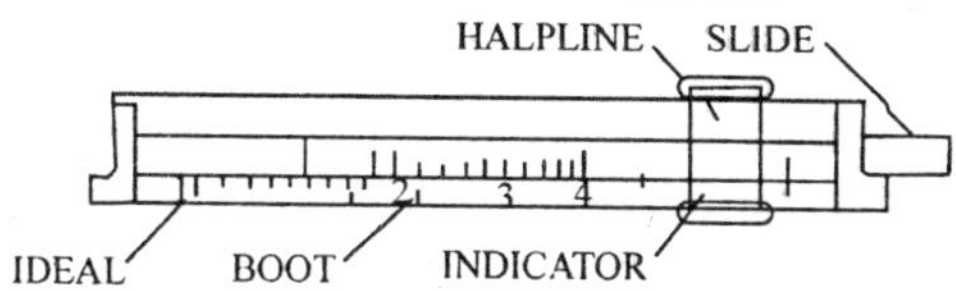

Fig. Simplified Diagram of a slide rule an Index at the Extreme left and at the Extreme right on both the C and D scales

SLIDE RULE THEORY

We have mentioned that the slide rule is based on logarithms. Recall that, to multiply two numbers, we simply add their logarithms. Previously we found these logarithms in tables, but if the logarithms are laid off on scales such as the C and D scale of the slide rule, we can add the lengths, which represent these logarithms. To make such a scale we could mark off mantissas ranging from 0 to 1 on a rule as in figure. We then find in the tables the logarithms for numbers ranging from 1 to 10 and write the number opposite its corresponding logarithm on the scale.

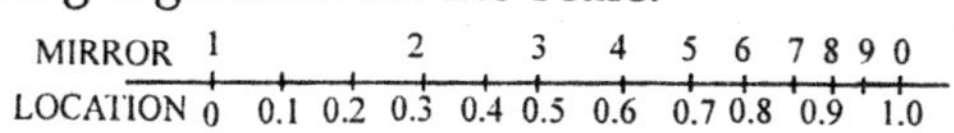

Fig. Logarithms and Corresponding Numbers on a Scale

Table lists the numbers 1 through 10 and their corresponding logarithms to three places. These numbers are written opposite their logarithms on the scale shown in figure. If we have two such scales, exactly alike, arranged so that one of them is free to slide along the other, we can perform the operation of multiplication, for example, by Adding Lengths that is, by adding logarithms. For example, if we wish to multiply 2 x 3, we find the logarithm of 2 on the stationary scale and move the sliding scale so that its index is over that mark.

We then add the logarithm of 3 by finding that logarithm on the sliding scale and by reading below it, on the stationary scale, the logarithm that is the sum of the two. Since we are not interested in the logarithms themselves, but rather in the numbers they represent, it is possible to remove the logarithmic notation on the scale in figure, and leave only the logarithmically spaced number scale. The C and D scales of the ordinary slide rule are made up in this manner. Figure shows the multiplication of 2 x 3. Although the logarithm scales have been removed, the numbers 2 and 3 in reality signify the logarithms of

Table: Numbers and Their Corresponding Logarithms

Number	Logarithm	Number	Logarithm
1	0.000	6	0.778
2	0.301	7	0.845
3	0.477	8	0.903
4	0.602	9	0.945
5	0.699	10	1.000

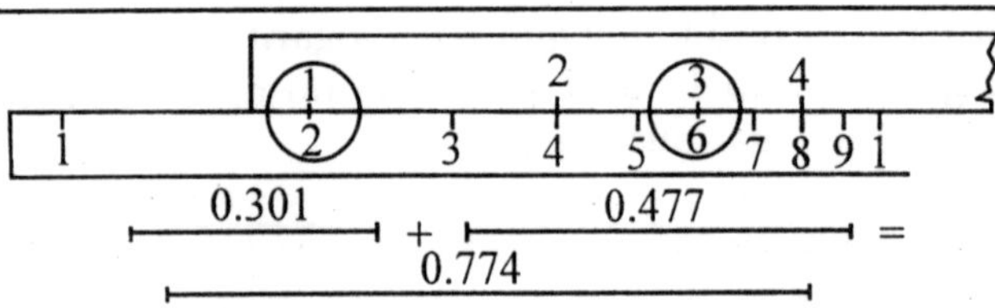

Fig. Multiplication by use of the slide rule

2 and 3, namely, 0.301 and 0.477; the product 6 on the scale really signifies the logarithm of 6, that is, 0.778. Thus, although

logarithms are the underlying principle, we are able to work with the numbers directly.

It should be noted that the scale is made up from mantissas only. The characteristic must be determined separately as in the case where tables are used. Since mantissas identify only the digit Mannheim sequence, the digit 3 on the slide rule represents not only 3 but 30, 300, 0.003, 0.3, and so forth.

Thus, the divisions may represent the number multiplied or divided by any power of 10. This is true also for numbers that fall between the divisions. The digit sequence, 1001, could represent 100.1, 1.001, 0.01001, and so forth. The following example shows the use of the same set of mantissas which appear in the foregoing example, but with a different characteristic and, therefore, a different answer:

Example: Use logs (positions on the slide rule) to multiply 20 times 30.

Solution:

log 20 = 1.301 (2 on the slide rule)

log 30 = 1.477 (3 on the slide rule)

log of answer = 2.778 (6 on the slide rule)

Since the 2 in the log of the answer is merely the indicator of the position of the decimal point in the answer itself, we do not expect to find it on the slide rule scale. As in the foregoing example, we find the digit 6 opposite the multiplier 3. This time, however, the 6 represents 600, because the characteristic of the log represented by 6 in this problem is 2.

READING THE SCALES

Reading a slide rule is no more complicated than reading a yard stick or ruler, if the differences in its markings are understood. Between the two indices of the C or D scales (the large digit 1 at the extreme left and right of the scales) are divisions numbered 2, 3, 4,5, 6, 7,8, and 9. Each length between two consecutive divisions is divided into 10 sections and each section is divided into spaces.

Notice that the division between 1 and 2 occupies about one-third of the length of the rule. This is sufficient space in which to write a number for each of the section marks. The sections in the remaining divisions are not numbered, because the space is more limited. Notice also that in the division between 1 and 2, the sections are each divided into 10 spaces.

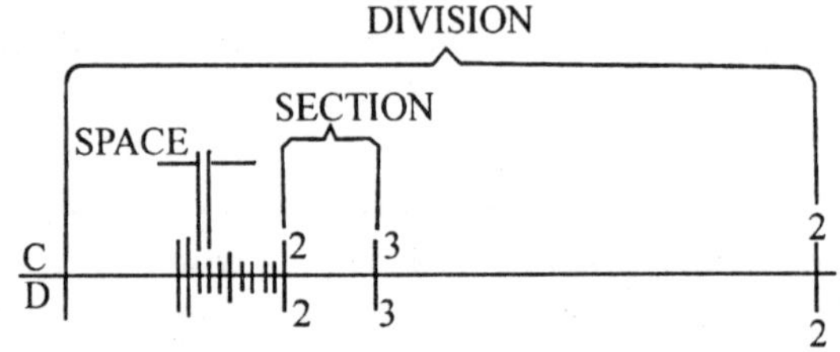

Fig. Division, Section, and Space of a Slide Rule Scale

The sections of the divisions from 2 to 4 are subdivided into only 5 spaces, and those from 4 to the right index are subdivided into only 2 spaces. These subdivisions are so arranged because of the limits of space.

Only the sequence of significant digits is read on the slide rule. The position of the decimal point is determined separately. For example, if the hairline of the indicator is in the left-hand position shown in figure, the significant digits are read as follows:

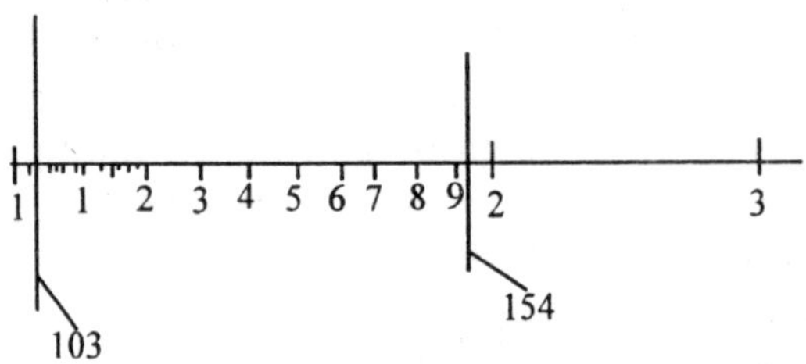

Fig. Readings in the first Division of a Slide Rule

1. Any time the hairline falls in the first division, the first significant digit is 1.
2. Since the hairline lies between the index and the first section mark, we know the number lies between 1.0 and 1.1, or 10 and 11, or 100 and 110, etc. The second significant digit is 0.
3. We next find how far from the index the hairline is located. It lies on the marking for the third space.
4. The three significant digits are 103. In the second

example shown in figure the hairline is located in the first division, the ninth section, and on the fourth space mark of that section. Therefore, the significant digits are 194. number from 0 through 9 as its second digit and any number from 0 through 9 as its third digit. Sometimes a fourth digit can be roughly approximated in this first division, but the number is really accurate to only three significant digits.

In the second and third divisions, each section is divided into only 5 spaces. Thus, each space is equal to 0.2 of the section. Suppose, for example, that the hairline lies on the third space mark after the large 2 indicating the second division. The first significant digit is 2. Since the hairline lies between 2 and the first section mark, the second digit is 0. The hairline lies on the third space mark or 0.6 of the way between the division mark and the first section mark, so the third digit is 6. Thus, the significant digits are 206. Notice that if the hairline lies on a space mark the third digit can be written accurately; otherwise it must be approximated. Thus, we see that any number falling in the first division of the slide rule will always have 1 as its first significant digit. It can have any From the fourth division to the right index, each section is divided into only.two spaces.

Thus, if the hairline is in the fourth division and lies on the space mark between the sixth and seventh sections, we would read 465. If the hairline did not fall on a space mark, the third digit would have to be approximated.

OPERATIONS WITH THE SLIDE RULE

There are two parts in solving problems with' a slide rule. In the first part the slide rule is used to find the digit sequence of the final result. The second part is concerned with the placing of the decimal point in the result. Let us consider first the digit sequence in multiplication and division.

Multiplication

Multiplication is performed on the C and D scales of the slide rule.

The following procedure is used:

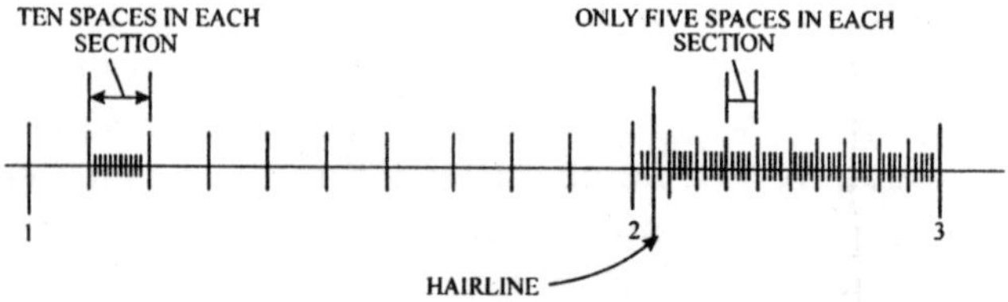

Fig. Reading in the Second Division of a Slide Rule

1. Locate one of the factors to be multiplied on the D scale, disregarding the decimal point.
2. Place the index of the C scale opposite that number.
3. Locate the other factor on the C scale and move the hairline of the indicator to cover this factor.
4. The.product is on the D scale under the hairline.

Sometimes in multiplying numbers, such as 25 x 6, the number on the C scale extends to the right of the stock and the product cannot be read. In such a case, we simply shift indices. Instead of the left-hand index of the C scale, the right-hand index is placed opposite the factor on the D scale. The rest of the problem remains the same. By shifting indices, we are simply multiplying or dividing by 10, but this plays no part in reading the significant digits. Shifting indices affects the characteristic only.

Example: 252 × 3 = 756

1. Place the left index of the C scale over 252.
2. Locate 3 on the C scale and set the hair- line of the indicator over it.
3. Under the hairline on the D scale read the product, 756.

Example: 4 × 64 = 256

1. Place the right index of the C scale over 4.
2. Locate 64 on the C scale and set the hairline of the indicator over it.
3. Under the hairline on the D scale read the product, 256.

Division

Division being the inverse of multiplication, the process

of multiplication is reversed to perform division on a slide rule. We subtract the length representing the logarithm of the divisor from the length representing the logarithm of the dividend to get the logarithm of the quotient.

The procedure is as follows:

1. Locate the dividend on the D scale and place the hairline of the indicator over it.
2. Move the slide until the divisor (on the C scale) lies under the hairline.
3. Read the quotient on the D scale opposite the C scale index.

If the divisor is greater numerically than the dividend, the slide will extend to the left. If the divisor is less, the slide will extend to the right. In either case, the quotient is the number on the D scale that lies opposite the C scale index, falling within the limits of the D scale.

Example: 6 ÷ 3 = 2

1. Locate 6 on the D scale and place the hairline of the indicator over it.
2. Move the slide until 3 on the C scale is under the hairline.
3. Opposite the left C scale index, read the quotient, 2, on the D scale.

Example: 378 ÷ 63 = 6

1. Locate 378 on the D scale and move the hairline of the indicator over it.
2. Move the slide to the left until 63 on the C scale is under the hairline.
3. Opposite the right-hand index of the C scale, read the quotient, 6, on the D scale.

PLACING THE DECIMAL POINT

Various methods have been advanced regarding the placement of the decimal point in numbers derived from slide rule computations. Probably the most universal and most easily remembered method is that of approximation.

The method of approximation means simply the rounding off of numbers and the mechanical shifting of decimal points in the numbers of the problem so that the approximate size of the solution and the exact position of the decimal point will be seen from inspection.

The slide rule may then be used to derive the correct sequence of significant digits. The method may best be demonstrated by a few examples. Remember, shifting the decimal point in a number one place to the left is the same as dividing by 10. Shifting it one place to the right is the same as multiplying by 10. Every shift must be compensated for in order for the solution to be correct.

Example: 0.573 × 1.45

Solution: No shifting of decimals is necessary here. We see that approximately 0.6 is to be multiplied by approximately 1 1/2. Immediately, we see that the solution is in the neighbourhood of 0.9. By slide rule we find that the significant digit sequence of the product is 832. From our approximation we know that the decimal point is to the immediate left of the first significant digit, 8. Thus,

$$0.573 \times 1.45 = 0.832$$

Example: 239 × 52.3

Solution: For ease in multiplying, we shift the decimal point in 52.3 one place to the left, making it 5.23. To compensate, the decimal point is shifted to the right one place in the other factor. The new position of the decimal point is indicated by the presence of the caret symbol.

$$239.0_{\wedge} \times 5_{\wedge}2.3$$

Our problem is approximately the same as

$$2{,}400 \times 5 = 12{,}000$$

By slide rule the digit sequence is 125. Thus,

$$239 \times 52.3 = 12{,}500$$

Example: 0.000134 × 0.092

Solution: Shifting decimal points, we have

$$0_{\wedge}00.000134 \times 0.09_{\wedge}2$$

Approximation: 9 × 0.0000013 = 0.0000117. By slide rule the digit sequence is 123. From approximation the decimal point is located as follows: 0.0000123

Thus, 0.000134 × 0.092 = 0.0000123

Example: $\frac{53.1}{42.4}$

Solution: The decimal points are shifted so that the divisor becomes a number between 1 and 10. The method employed is cancellation. Shifting decimal points, we have

$$\frac{5 \wedge 3.1}{4 \wedge 2.4}$$

Approximation: 5/4 = 1.2

Digit sequence by slide rule:

1255

Placing the decimal point from the approximation:

1.255

Thus,

$$\frac{53.1}{42.4} = 1.255$$

Example:

$$\frac{0.00645}{0.0935}$$

Solution: Shifting decimal points

$$\frac{0.00 \wedge 645}{0.09 \wedge 35}$$

Approximation:

$$\frac{0.6}{9} = 0.07$$

Digit sequence by slide rule: 690

Placing the decimal point from the approximation:

0.0690

Thus,

$$\frac{0.00645}{0.0935} = 0.0695$$

MULTIPLICATION AND DIVISION COMBINED

In problems such as,

$$\frac{0.644 \times 330}{161 \times 12}$$

It is generally best to determine the position of the decimal point by means of the method of approximation and to determine the significant digit sequence from the slide rule. Such problems are usually solved by dividing and multiplying alternately throughout the problem. That is, we divide 0.644 by 161, multiply the quotient by 330, and divide that product by 12.

Shifting decimal points, we have,

$$\frac{0 \wedge 0644 \times 3 \wedge 30}{1 \wedge 61 \times 1 \wedge 2}$$

Since there is a combined shift of three places to the left in the divisor, there must also be a combined shift of three places to the left in the dividend.

Approximation: $\frac{0.06 \times \not{3}^{2}}{\not{1}.\not{5}} = 0.06 \times 2 = 0.12.$

The step-by-step process of determining the significant digit sequence of this problem is as follows:

1. Place the hairline over 644 on the D scale.
2. Draw the slide so that 161 of the C scale lies under the hairline opposite 644.
3. Opposite the C scale index (on the D scale) is the quotient of 644 + 161. This is to be multiplied by 330, but 330 projects beyond the rule so the C scale indices must be shifted.
4. After shifting the indices, find 330 on the C scale and place the hairline over it. Opposite 330 under the hairline on the D scale is the product of,
$$\frac{644}{161} \times 330.$$
5. Next, move the C scale until 12 is under the hairline. Opposite the C scale index (on the D scale) is the final quotient. The digit sequence is 110.

The decimal point is then placed according to our approximation: 0.11. Thus,

$$\frac{0.644 \times 330}{161 \times 12} = 0.11.$$

SQUARES

Squares of numbers are found by reference to the A scale.

The numbers on the A scale are the squares of those on the D Scale. The A scale is really a double scale, each division being one-half as large as the corresponding division on the D scale. The use of a double scale for squaring is based upon the fact that the logarithm of the square of a number is twice as large as the logarithm of the number itself.

In other words,

$$\log N^2 = 2 \log N$$

This is reasonable, since,

$$\log N^2 = \log (N \times N) = \log N + \log N$$

For a numerical example, suppose that we seek to square 2 by means of logarithms.

$$\log 2 = 0.301$$

$$\log 2^2 = 2 \log 2 = 2 \times 0.301 = 0.602$$

Since each part of the A scale is half as large as the corresponding part of the D scale, the logarithm 0.602 on the A scale will be the same length as the logarithm 0.301 on the D scale. That is, these logarithms will be opposite on the A and D scales. On the A scale as on the D scale, the numbers are written rather than their logarithms. Select several numbers on the D scale, such as 2, 4, 8, 11, and read their squares on the A scale, namely 4, 16, 64, 121.

Notice also that the same relation exists for the B and C scales as for the A and D scales. Of interest, also, is the fact that since the A and B scales are made up as are the C and D scales, they too could be used for multiplying or dividing.

Placing the Decimal Point

Usually the decimal may be placed by the method of approximation. However, close observation will reveal certain facts that eliminate the need for approximations in squaring numbers.

Two rules suffice for squaring whole or mixed numbers, as follows:

1. When the square of a number is read on the left half of the A scale, that number will contain twice the number of digits to the left of the decimal point in the original number, less 1.

2. When the square of a number is read on the right half of the A scale, that number will contain twice the number of digits to the left of the decimal point in the original number.

Example: Square 2.5.

Solution: Place the hairline over 25 on the D scale. Read the digit sequence, 625, under the hairline in the left half of the A scale.

By rule 1: (2xnumberof digits)–1 = 2(1)–1 = 1. There is one digit to the left of the decimal point. Thus,

$$(2.5)^2 = 6.25$$

Example: Square 6,340.

Solution: Digit sequence, right half A scale: 402.

By rule 2: 2 × number of digits = 2 × 4 = 8 (digits in answer). Thus,

$$(6{,}340)^2 = 40{,}200{,}000$$

Positive Numbers Less Than One

If positive numbers less than one are to be squared, a slightly different version of the preceding rules must be employed. Count the zeros between the decimal point and the first nonzero digit. Consider this count negative. Then the number of zeros between the decimal point and the first significant digit of the squared number may be found as follows:

1. *Left half A scale*: Multiply the zeros counted by 2 and subtract 1.
2. *Right half A scale*: Multiply the zeros counted by 2.

Example: Square 0.0045

Solution: Digit sequence, right half A scale: 2025. By rule 2: 2(–2) = –4. (Thus, 4 zeros between the decimal point and the first digit.)

$$(0.0045)^2 = 0.09002025$$

Example: Square 0.0215

Solution: Digit sequence, left half A scale: 462. By rule 1: 2(–1) –1 = –3

$$(0.0215)^2 = 0.000462$$

SQUARE ROOTS

Taking the square root of a number with the slide rule is the inverse process of squaring a number. We find the number on the A scale, set the hairline of the indicator over it, and read the square root on the D scale under the hairline.

Positioning Numbers on the A Scale Since there are two parts of the A scale exactly alike and the digit sequence could be found on either part, a question arises as to which section to use. Generally, we think of the left half of the rule as being numbered from 1 to 10 and the right half as being numbered from 10 to 100. The numbering continues- left half 100 to 1,000, right half 1,000 to 10,000, and so forth.

A simple process provides a check of the location of the number from which the root is to be taken. For whole or mixed numbers, begin at the decimal point of the number and mark off the digits to the left (including end zeros) in groups of two.

This is illustrated in the following two examples:

1. $\sqrt{40,300.21}$ $\sqrt{4'03'00.21}$
2. $\sqrt{2.034.1}$ $\sqrt{20'34.1}$

Look at the left-hand group. If it is a l-digit number, use the left half of the A scale. If it is a a-digit number, use the right half of the A scale. The number in example 1 is thus located in the left half of the A scale and the number in example 2 is located in the right half.

Numbers Less Than One

For positive numbers less than one, begin at the decimal point and mark off groups of two to the right. This is illustrated as follows:

1. $\sqrt{0.000245}$ $\sqrt{0.00'02'45}$
2. $\sqrt{0.00402}$ $\sqrt{0.00'40'2}$

Looking from left to right, locate the first group that contains a digit other than zero. If the first figure in this group it3 zero, locate the number in the left half of the A scale. If the first figure is other than zero, locate the number in the right half of the A scale. Thus, $\sqrt{0.00'02'45}$ is located left. and $\sqrt{0.00'40'2}$ is located right.

Powers of 10

When the square root of 10, 1,000, 100,000, and so forth, is desired, the center index is used. That is, when the number of digits in a power of 10 is even, use the center index. The slide rule uses only the first three significant digits of a number. Thus, if the rule is used, $\sqrt{23451.6}$ must be considered as $\sqrt{23500.0}$. Likewise, 1.43567 would be considered 1.43000, and so forth. For greater accuracy, other methods must be used.

Placing the Decimal Point

To place the decimal point in the square root of a number, mark off the original number in groups of two as explained previously. For whole or mixed numbers, the number of groups marked off is the number of digits including end zeros to the left of the decimal point in the root.

The following problems illustrate this:

1. $\sqrt{23,415}$, $\sqrt{2'34'15}$: Three digits to left of decimal point in square root
2. $\sqrt{421,562.4}\sqrt{42'15'62.4}$: Three digits to left of decimal point in square root
3. $\sqrt{231.321}\sqrt{2'31.321}$: Two digits to left of decimal point in square root

For positive numbers less than one, there will be one zero in the square root between the decimal point and the first significant digit for every pair of zeros counted between the decimal point and the first significant digit of the original number.

This is illustrated as follows:

1. $\sqrt{0.0004}\sqrt{0.00'04}$: One zero before first digit in square root
2. $\sqrt{0.00008}\sqrt{0.00'00'8}$: Two zeros before first digit in square root
3. $\sqrt{0.08'}$: No zeros before first digit in square root

Example: $\sqrt{4,521}\sqrt{45'21}$ (Two digits in left-hand group)

Place the hairline over 452 on the right half of the A scale. Read the digit sequence of the root, 672, on the D scale under the hairline. Since there are two groups in the original number, there are two digits to the left of the decimal point in the root.

Thus, $\sqrt{4,521} = 67.2$

Example: $\sqrt{0.000741}\sqrt{0.00'07'41}$: (First figure is zero in this group)

Place the hairline over 741 on the left half of the A scale. Read the digit sequence of the root, 272, under the hairline on the D scale. Since there is one pair of zeros to the left of the group containing the first digit, there is one zero between the decimal point and the first significant digit of the root.

Thus,

$$\sqrt{0.000741} = 0.0272$$

CUBES AND CUBE ROOTS

Cubes and cube roots are read on the K and D scales of the slide rule. On the K scale are compressed three complete logarithmic scales in the same space as that of the D scale. Thus, any logarithm on the K scale is three times the logarithm opposite it on the D scale. To cube a number by logarithms, we multiply its logarithm by three. Therefore, the logarithms of cubed numbers will lie on the K scale opposite the numbers on the D scale. As with the other slide rule scales mentioned, the numbers the logarithms represent, rather than the logarithmic notations, are printed on the rule. In the left-hand third of the K scale, the numbers range from 1 to 10; in the middle third they range from 10 to 100; and in the right-hand third, they range from 100 to 1,000. To cube a number, find the number on the D scale, place the hairline over it, and read the digit sequence of the cubed number on the K scale under the hairline.

Placing the Decimal Point

The decimal point of a cubed whole or mixed number may be easily placed by application of the following rules:

1. If the cubed number is located in the left third of the K scale, its number of digits to the left of the decimal point is 3 times the number of digits to the left of the decimal point in the original number, less 2.

2. If the cubed number is located in the middle third of the K scale, its number of digits is 3 times the number of digits of the original number, less 1.
3. If the cubed number is located in the right third of the K scale, its number of digits is 3 times the number of digits of the original number.

Example: $(1.6)^3$

Solution: Place the hairline over 16 on D scale. Read the digit sequence, 409, on the K scale under the hairline. Number of digits to left of decimal point in the number 1.6 is 1 and the cubed number is in the left-hand third of the K scale.

$$3 \times (\text{No. of digits}) - 2 = (3 \times 1) - 2 = 1$$

Therefore,

$$(1.6)^3 = 4.09$$

Example: $(4.1)^3$

Digit sequence = 689.

Solution: Number of digits to left of decimal point in the number 4.1 is 1, and the cubed number is in the middle third of the K scale.

$$3 \times (\text{No. of digits}) - 1 = (3 \times 1) - 1 = 2$$

Therefore,

$$(4.1)^3 = 68.9$$

Example: $(52)^3$

Solution: Digit sequence = 141.

Number of digits to left of decimal point in the number 52 is 2, and the cubed number is in the right-hand third of the K scale.

$$3 \times \text{No. of digits} = 3 \times 2 = 6$$

Therefore,

$$(52)^3 = 141{,}000$$

Positive Numbers Less Than One

If positive numbers less than one are to be cubed, count the zeros between the decimal point and the first nonzero digit. Consider the count negative. Then the number of zeros between the decimal point and the first significant digit of the cubed number may be found as follows:

1. Left third of K scale: Multiply the zeros counted by 3 and subtract 2.
2. Middle third of K scale: Multiply the zeros counted by 3 and subtract 1.
3. Right third of K scale: Multiply the zeros counted by 3.

Example: Cube 0.034

Solution: Digit sequence = 393

Zero count of 0.034 = –1, and 393 is in the middle third of the K scale. 3 × (No. of zeros)–1 = (3 × –1)–1 = –4

Therefore,

$$(0.034)^3 = 0.0000393$$

Cube Roots

Taking the cube root of a number on the slide rule is the inverse process of cubing a number. To take the cube root of a number, find the number on the K scale, set the hairline over it, and read the cube root on the D scale under the hairline.

POSITIONING NUMBERS ON THE K SCALE

Since a given number can be located in three positions on the K scale, the question arises as to which third of the K scale to use when locating a number. Generally, the left index, the left middle index, the right middle index, and the right index are considered to be numbered as shown in figure.

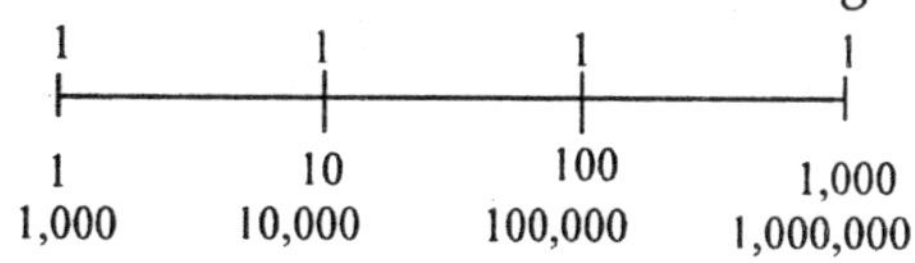

Fig. Powers of 10 Associated with K-scale indices.

A system similar to that used with square roots may be used to locate the position of a number on the K scale. Groups of three are used rather than groups of two.

The grouping for cube root is illustrated as follows:

1. $\sqrt[3]{40,531.6}$, $\sqrt[3]{40'531.6}$
2. $\sqrt[3]{4,581.43}$, $\sqrt[3]{4'561.43}$
3. $\sqrt[3]{0.000043}$, $\sqrt[3]{0.000'043}$

For whole or mixed numbers the following rules apply:

1. If the left-hand group contains one digit, locate the number in the left third of the K scale.
2. If the left group contains two digits, locate the number in the middle third of the K scale.
3. If the left group contains three digits, locate the number in the right third of the K scale.

The following examples illustrate the foregoing rules:

1. $\sqrt[3]{4'561.43}$
   (One digit)-left third K scale.
2. $\sqrt[3]{40'531.6}$
   (Two digits)-middle third K scale.
3. $\sqrt[3]{453'361}$
   (Three digits)-right third of K scale.

For positive numbers less than one, look from left to right and find the first group that contains a digit other than zero.

1. If the first two figures in this group are zeros, locate the number in the left third of the K scale.
2. If only the first figure in this group is zero, locate the number in the middle third of the K scale.
3. If the first figure of the group is not zero, locate the number in the right third of the K scale.

The following examples illustrate these rules:

1. $\sqrt[3]{0.000'004'53}$
   (Two zeros)-left third K scale.
2. $\sqrt[3]{0.000'050'43}$
   (One zero)-middle third K scale.
3. $\sqrt[3]{0.000'000'430}$
   (No zero)-right third K scale.

PLACING THE DECIMAL POINT

To place the decimal point in the cube root of a number, we use the system of marking off in groups of three as shown above. For whole or mixed numbers, there is one digit in the root to the left of the decimal point for every group marked in the original number. Thus,

$$\sqrt[3]{4'531.6}$$

(Two digits in root to left of decimal point.)

For positive numbers less than one, there will be one zero in the root between the decimal point and the first significant digit for every three zeros counted between the decimal point and the first significant digit of the original number. Thus,

$$\sqrt[3]{0.000'000'004}$$

(Two zeros between decimal point and first significant digit of root.)

Example: $\sqrt[3]{216000.4}$, $\sqrt[3]{216'000.4}$

(Three digits in left group)

Place the hairline over 216 in the right third of the K scale. Read the digit sequence, 6, under the hairline on the D scale.

Since there are two groups in the original number, there are two digits to the left of the decimal point in the root.

Thus,

$$\sqrt[3]{216000.4} = 60$$

Example: $\sqrt[3]{0.0000451} = \sqrt[3]{0.000'045'1}$

(Only first figure is zero in this group) Place the hairline over 451 in the middle third of the K scale. Read the digit sequence, 357, under the hairline on the D scale. Since there is one group of three zeros, there is one zero between the decimal point and the first significant digit of the root.

Thus,

$$\sqrt[3]{0.0000451} = 0.0357$$

POWERS OF 10

To take the cube root of a power of 10, mark it off as explained in the preceding paragraphs. The number in the left group will then be 1, 10, or 100. We know that the cube root of 10 is a number between 2 and 3. Thus, for the cube root of any number whose left group is 10, use the K scale index which lies between 2 and 3 on the D scale. The cube root of 100 lies between 4 and 5. Therefore, for a number whose left group is 100, use the K scale index between 4 and 5 on the D scale.

2

The Arithmetic Mean

Suppose a series of magnitudes, x_1, x_2, x_3..., x_N, all recognized as belonging to the same variable, to have been recorded in quantitative description. If all these individual values are identical, it is not necessary to determine mathematically an expression to represent them. The value required would be obvious by inspection. Any one value of x would be entirely representative of the whole series.

Complete concordance of all values of a variable is rarely, if ever, encountered in biological work. A situation of perfect concordance would (or should) immediately incite suspicion that the observations are biased. Indeed, in all fields of science, as the instruments for measurement become more refined it is increasingly apparent that things which may at first have seemed identical really differ among themselves.

Under such conditions no single individual measurement can be accepted as being wholly representative of the entire sample. It becomes necessary to obtain a description of the series as a whole that will take into account all the measures available. A first step is to establish a typical value for the variate magnitudes—a single value which will be representative of the series of magnitudes in some specified and appropriate manner.

THE ARITHMETIC MEAN

The average or mean value is so familiar as a typical magnitude that it seems superfluous to discuss its functions; and yet, like much that is commonplace, its philosophical implications rarely receive consideration. It provides a measure of type, the usefulness and appeal of which are attested by the extensive service which it has rendered since calculation first began. It is simple both to comprehend and to compute. It is perfectly impartial and unselective, for every observation enters into its magnitude without any trace of distortion. All who have employed it—and who, in fact, has not?—have in truth ascended to the first rung of the ladder of statistical analysis. A very useful symbol for the mean value of a variable is provided by placing a bar above the letter designating the variable. If x is that variable, then the symbol for the mean, x+ (customarily verbalized as "x bar"), finds some preference over such alternatives as M_x (mean of x) or A_x (average of x) because of its avoidance of the subscript notation. In algebraic definition of x+,one may write,

$$\bar{x} = \frac{x_1 + x_2 + x_3 + \cdots + x_N}{N}$$

$$= \frac{\sum x}{N}$$

The Greek letter Σ (pronounced "sigma"), which is the equivalent of the English capital letter S, denotes summation for the entire series of values indicated by the symbols that immediately follow it. It is merely a mathematical shorthand symbol for continued addition of all values corresponding to the one given. When the equation is rearranged to the form,

$$\bar{x} = \frac{x_1}{N} + \frac{x_2}{N} + \frac{x_3}{N} + \cdots + \frac{x_N}{N}$$

$$= \frac{\sum x}{N}$$

The mean appears as a sum compiled from one Nth part of every magnitude considered. Each individual in the series is therefore represented in the mean in proportion to its magnitude. This property defines the sense in which to the lay mind the mean logically typifies any series for which it is calculated.

Technically, equation defines the arithmetic mean of a series. Two other types of average generated by the arithmetic mean on transformed scales and therefore appropriate to those special conditions justifying the transformation are the geometric and harmonic means, defined respectively by G and H in the equations:

$$\log G = \frac{\Sigma(\log x)}{N}, \text{ or } G = \sqrt[N]{x_1 x_2 \cdots x_N}$$

$$H^{-1} = \frac{\Sigma(x^{-1})}{N}, \text{ or } \frac{1}{H} = \frac{1}{N}\left[\frac{1}{x_1} + \frac{1}{x_2} + \cdots + \frac{1}{x_N}\right]$$

Since these statistics pertain to special types of analysis of relatively rare occurrence in biometric work, further discussion of them will not be given here. When the term *mean* is used without further designation, the arithmetic mean is always implied. A fundamentally different approach to the problem of establishing a representative value for a series of variates may be made through selection of some particular variate or group of variates as typical of the whole. Two such selections came into rather common usage in the days when calculating machines were very much more of a luxury than they are now. These two descriptive values became more or less obvious as a result of organization of the data and do not depend primarily on calculation. They merit careful consideration at the present time, not as short-cut approximations to the mean for which purpose they appear to have been widely used, but as quantities descriptive of type in different senses from the mean. One may perhaps more readily grasp the descriptive value of these quantities through derivation of them directly from actual data.

THE MEDIAN

In a study of the erythrocyte count (red blood cell count) for normal men, Haden gives findings for 40 individuals. Arranged in order of magnitude, these counts were as follows:

4.27,	4.32,	4.40,	4.52,	4.56,	4.58,	4.64,	4.70,	4.72,	4.73,
4.80,	4.80,	4.80,	4.80,	4.84,	4.87,	4.89,	4.93,	4.97,	4.98,
4.99,	5.00,	5.02,	5.05,	5.09,	5.09,	5.10,	5.15,	5.16,	5.20,
5.20,	5.20,	5.26,	5.28,	5.36,	5.46,	5.49,	5.50,	5.57,	5.62,

millions per cubic millimeter

This arrangement of a series of values in order of magnitude is known as *ranking*. It enables one to determine quickly how many individuals fall below and above any given magnitude. The problem is to select a magnitude which is representative of the set. One may endeavor to do this with the idea of a central value in mind. Since the above series has an even number of variates, there is no "middlemost" individual. However, 20 individuals have a count not exceeding 4.98, and 20 have a count not less than 4.99.

The mid-point between these two magnitudes may logically be selected as the middlemost magnitude. It is known technically as the median value. Had N been odd, then there would have been a middlemost individual, giving a unique solution to the problem of ascertaining the value above and below which an equal number of observations occur. Of course the situation would become complicated again if several variates had that same magnitude, but we need not pause to argue this point.

The median may be accepted as being representative in the sense that it divides the total frequency into halves; it is central with respect to the distribution of the total frequency along the scale.

As the central value with respect to frequency distribution, the median has the peculiar property of being unchanged by alteration of any or all of the other variate magnitudes in the series, provided that the *sign* of deviation from the median remains unchanged in each case. Stated inversely, the numerical magnitudes of the individual deviations from the median in a sample do not influence the median for that sample.

Thus the median is often quite useful as a measure of "central tendency" in a sample where it is desired to suppress the influence of the extreme or unusual variate magnitudes. It is a magnitude representative of mediocrity, serving to eliminate the influence of the geniuses and the morons for the student of educational statistics, or the influence of "epidemics" and periods of unusual good health for the vital statistician, and so forth.

THE MODAL VALUE

The organization of a series of data by ranking the variates places identical magnitudes together and automatically draws attention to the relative frequencies of occurrence of the observed values. It is a logical stepping stone to the form of organization which we have already considered under the title of seriation. Just as ranking naturally centers attention on the number of observations above and below any specified point, so seriation emphasizes the importance of successive values with respect to frequency of occurrence.

The class of greatest frequency naturally draws attention on this account, if the class ranges are uniform. It is called the modal or "most fashionable" class, the class center being commonly designated as the modal value. This provides immediately a value which is typical of the observations in a new and quite appropriate sense.

It is well to recognize at once that establishment of the modal value is fundamentally associated with a reasonably satisfactory definition of this point in the appropriate law of frequency distribution for the variable considered. If any seriation does not lead to a reasonably smooth frequency distribution, the question arises at once of the acceptability of the indicated modal value.

Now "smoothness" of a histogram is a function of size of sample and the unit of classification. This unit is chosen quite arbitrarily, and the indicated modal value may be shifted substantially according to this selection, even when N is quite large.

The difficulties of selecting the appropriate modal value by seriation of a small sample are often overwhelming. Let us look for a moment at the series of 40 erythrocyte counts above. The count of 4.80 occurs four times and is the most frequent individual value as the data are given. But what would one do about selecting the modal value if a recheck of the counts showed an error, one of the 4.80 values being changed, say, to 4.81? In that case both of the counts 4.80 and 5.20 would occur with a frequency of 3, giving two modal values rather widely separated.

Coarser grouping may be resorted to, but the indicated modal value will be found to vary around 5.00 according to the seriation scheme adopted. Study of the histograms for the same series of observations in the successive panels will quickly substantiate the general principle. As applied to a small sample, the modal value has rather poor descriptive capacity. From what has been said in the preceding chapter it is easy to see that this deficiency will diminish as the size of the sample is increased and its frequency distribution accordingly reflects more clearly the supply form.

Desirable Qualities of a Descriptive Statistic

These preliminary definitions of the mean, median, and modal values as independent and differing descriptions of typical value raise the question of the relative quality of alternative descriptions. It has been stated that quantitative description of the characteristic features of sets of data comprises a basic task in statistical analysis.

Numerical quantities must be devised which will describe these features accurately, so that comparisons of them in different variables may be readily accomplished in an objective manner. The term statistic is now commonly being applied to any summarizing quantity such as a mean, median, et cetera, derived from and descriptive of a set of variates. Using the term in this sense we may now well inquire as to the desirable qualities of good descriptive statistics in general.

- They should be based upon every variate in the sample. Assuming the basic data to be accurate, surely each measurement is just as important as any other measurement in defining the variable.
- They should be capable of algebraic definition. Algebra may be considered as providing the vehicle of symbolic definition for the development of logical thought sequences applicable to magnitudes differing by finite amounts. The utility of any statistic as a factor in such logical development of thought will depend on the generality of its algebraic definition.

- They should be readily comprehensible with respect to each descriptive function.
- They should be susceptible as far as possible of rapid determination.
- They should yield highly consistent values when applied to samples randomly drawn from the same supply.

While the mean fulfills every one of these conditions remarkably well, the same cannot be said of the modal and median values with equal validity. The latter are not capable of algebraic definition in general terms, nor do they depend directly for their values on the magnitudes of every variate in the series. Also, the consistency of means in random samples from the same supply is found to be considerably greater than that of median or modal values. If these three statistics are to be considered as alternatives in description, there can be no question as to the superiority of the mean. It should not be overlooked, however, that each has arisen as the result of a different point of view in approaching the problem of establishing a representative value. The problems in which they assume quite different magnitudes must be regarded as problems in which they are not alternative, each having its own descriptive function.

"Centering Values" of Frequency Curves

The preceding syntheses of the mean, median, and modal values as representations of typical magnitude referred to the law of frequency distribution only in the case of the modal value. The descriptive quality of the modal value was found to be very poor with small samples, but increased as refinement of seriation could proceed without loss of the form of the frequency distribution, that is, as the sample increased in size. It is only in the limit of a frequency curve being reached that the point of greatest frequency becomes independent of the various seriations to which the data may be subjected. It was urged by Karl Pearson that the term *mode* be reserved for the scale value at which the greatest ordinate occurs for the fitted frequency curve. Our previous use of the variant term *modal class* for the range within which maximum frequency occurs in a seriation was made with this distinction in mind.

When a histogram is graduated by a frequency curve, no attempt is made to force the mode of the curve to agree with the modal class of the histogram. The constants of the equation used fix the point at which the maximum ordinate will occur. These constants are determined from general properties of the observed data and are essentially independent of any seriation made. The curve given in panel (*D*). Must logically graduate each of the seriations given equally well. The various modal values of the several seriations contrast sharply with the fixed mode of the curve at 7.00 pounds.

The mean and median of a frequency curve are established precisely in accordance with the definitions already drawn for the variates. The median becomes the point on the scale of measurement at which the ordinate cuts the frequency curve into two segments of equal area, the latter—as we have sought to make clear—defining the frequency. It will be shown that the algebraic definition of the mean (equation) fixes it as the "point of balance" of a frequency system. That is, if a frequency distribution bounded apically by a curve is cut out of cardboard of uniform thickness and is balanced horizontally about an ordinate, then that ordinate will emanate from the base line at the mean value. It follows directly then that positive and negative deviations of the variates about any point in the scale of measurement of a variable balance one another *in toto* only when that point is taken at the mean of the distribution. This defines another attractive quality of the mean as a representative value.

For symmetrical frequency curves, the median, the mode, and the mean all coincide, as must surely be obvious. In a skew curve this is not so. The point of balance (the mean) must necessarily be drawn increasingly away from the point of greatest frequency (the mode) towards the long "tail" of the distribution as skewness intensifies. The amount of separation between them accordingly forms an excellent basis for quantitative description of the degree of skewness present in a frequency distribution. The median of non-symmetrical shaped curves falls between the mode and the mean, somewhat closer to the latter than the former. For small degrees of skewness it holds very closely to the approximating equation

(mode-mean)(=) 3 (median-mean).

Solving this equation to provide quickly an estimate of the mode after mean and median have been determined from a set of observations is perhaps acceptable as a rough approximation only if the sample is a large one and skewness is not very marked. Rather extensive calculation is involved in accurate determination of the mode, for the equation to the graduating curve must be fully determined first. Because of this there is temptation to use the above approximation without proper regard to its limitations, a temptation to be avoided, of course. In that which follows in this book we shall assume, as indeed we must to progress satisfactorily, that simple computations present no obstacle. It will be shown in the following section that a powerful device for the simplification of calculations may be mastered easily, placing the bulk of statistical work within the scope of elementary mental arithmetic. In view of this and the reasonable approach to symmetry commonly found in biological variables, the natural advantages of the mean as a typical value and reference point for the measurement of variation will cause its adoption in preference to the median for practically all work.

Coding as an Aid in Computation

The mean must be determined by calculation. Algebraically, this seems a most simple procedure, and indeed it is, provided

1. That the number of variates to be summed is not so large as to cause fatigue when handled individually, and
2. That the variate magnitudes are numbers of very few digits.

If either of these difficulties is present, they may readily be circumvented. The first may be avoided by seriation to form a frequency table. The second is very simply overcome by *coding,* which consists of replacing the true variate magnitudes by a simpler set of numbers according to a systematic scheme. The original scale of measurement may be replaced by an arbitrarily chosen scale of the simplest possible form in accordance with a well-known mathematical principle. All computations may be made in terms of the arbitrary scale and the results quickly transferred to their corresponding original scale values.

The objectives of seriation have already received attention in the pages. It may not be amiss at this point, however, to consider certain procedural details as they relate to the establishment of the computational table. Solving the following practical problem will serve as a medium for development of these points, as well as others incidental to the subject of coding.

In routine analysis of the protein content of grain arriving at a port, 1,937 determinations are made available. The protein content of each sample is expressed on a percentage basis to the nearest 0.05 per cent. The lowest value is 10.30 per cent, and the largest is 17.15 per cent, practically all intermediate values by increments of 0.05 occurring. How may the computation of the mean for such a series of values be most advantageously simplified?

The very large number of variates to be considered makes seriation a practical necessity. Since the increment of measurement is 0.05, there will be 138 possible values between the extreme limits of 10.30 and 17.15. Grouping into classes of sufficiently wide but uniform range to reduce the number of classes to about 12-16 would be desirable. A class range of 0.5 per cent protein would give approximately 14 classes. That range is so convenient for seriation purposes that it may well be chosen. "At what value should the beginning class boundary be set?" The reply to this question must be that, like selection of the class range, this matter remains entirely at the discretion of the individual. Since the class centers become the reference values after the seriation is made, some workers would aim to choose the boundaries so that the centers become simple numbers, like the even and half percentages.

This selection would place the class boundaries at values ending in.25 and.75, and the question would arise as to what to do with the large numbers of variates which fall precisely on these boundaries. Having the true class range an odd multiple of the increment of measurement avoids this difficulty, but, having due regard to ease of seriation, that would be inconvenient here, so we shall proceed with the suggested class range of 10 increments of measurement.

Having as simple a pattern of numbers as possible for the apparent class boundaries greatly facilitates seriation. Thus apparent class ranges of 10.0-10.45, 10.50-10.95, 11.0-11.45, et cetera, will lead to quicker seriation than possibly any other scheme with a true class range of 0.5 unit. We shall see later that it is quite immaterial to the calculations what the complexity of the class centers may happen to be, so we shall seriate according to this plan. Since the measurements were recorded to the "nearest" scale graduation, the mid-point of the apparent range for each class is identical with the center of the true class range. These centers are entered as column (3) of the table. For computational purposes they will replace the ranges, the effect of the seriation being to represent the data as if they had been recorded by increments of 0.5 per cent instead of 0.05.

Proceeding now to compute the mean value for this distribution, one must first multiply each class center, x, by the number of individuals in the class, f, to secure the sum of all variate magnitudes within the class. Column (4) contains these products, the grand total of the column being Σx for the entire distribution. By dividing this total by 1,937, the mean protein content is found to be 13.233 per cent.

The reader who has taken the trouble to verify the calculation of the mean, if he did not use a machine for multiplications, will have been impressed by the onerous mental arithmetic imposed by the five-digit numbers for the class centers. It is the particular function of coding to reduce this labour to a minimum through replacement of the existing set of class center or x values by a simpler set. In practice, coding is a most direct and elementary procedure, used in one form or another on occasion by all. It would seem advantageous first of all to trace the underlying steps with respect to a very simple illustration and then give full generality to the principle in algebraic form.

Table: A Frequency Distribution of Protein Analyses, with Calculations for the Mean Value

(Original determinations to the nearest 0.05 per cent protein)

Apparent class Range	Frequency	Class center	Class total
	f	*x*	*fx*
(1)	(2)	(3)	(4)
10.0-10.45	4	10.255	40.300
10.5-10.95	18	10.725	193.050
11.0-11.45	65	11.225	729.625
11.5-11.95	157	11.725	1,840.825
12.0-12.45	270	12.225	3,300.750
12.5-12.95	327	12.725	4,161.075
13.0-13.45	334	13.225	4,417.150
13.5-13.95	276	13.725	3,788.100
14.0-14.45	212	14.225	3,015.700
14.5-14.95	141	14.725	2,076.225
15.0-15.45	79	15.225	1,202.775
15.5-15.95	31	15.725	487.475
16.0-16.45	14	16.225	227.150
16.5-16.95	7	16.725	117.075
17.0-17.45	2	17.225	34.450
Total	**1937**		**25,632.325**

$\Sigma x = 25,632.325.$ $N = 1,937.$

$x + = 13.233.$

Probably any science student asked to give the average of the three numbers 19.01, 19.06, and 19.08 would immediately recognize that, since 19.0 is common to all, the average must be 19.0 followed by the average of 1, 6, and 8. He would be coding the data in so doing, finding the average on a new scale, then decoding his answer to get 19.05. Let us follow the steps in detail.

$$\frac{19.01+19.06+19.08}{3}=\frac{19+0.01+19+0.06+19+0.08}{3}$$

$$=\frac{3(19)}{3}+\frac{0.01+0.06+0.08}{3}$$

$$=19+\frac{0.01(1+6+8)}{3}$$

$$=19+0.01\left(\frac{15}{3}\right)$$

$$=19+0.01(5)$$

$$=19.05$$

"But," he might protest, "I didn't go through all that manipulation!" Perhaps what he should have said was that he did not realise that his short cuts were based on just that type of manipulation.

Let us now generalize these steps. Let x designate the original scale values, a be the constant 19, and b the constant 0.01. The numbers actually averaged (1, 6, and 8) were derived from the x values by subtracting a from each, then dividing the remainders by b. Designating the resulting code values by x, we have

$$x=\frac{x-a}{b} \text{ or } x = a + bx.$$

Now,

$$\bar{x}=\frac{\Sigma x}{N}=\frac{\Sigma(a+bx)}{N}$$

$$=\frac{\Sigma a}{N}+\frac{\Sigma bx}{N}=a+b\frac{\Sigma x}{N}=a+b\bar{x}$$

Therefore the mean value on the original scale of measurement may be found by securing the mean value in terms of a code scale and then decoding.

The reader who does not recall the algebraic rules of rearrangement about the summation sign, Σ, may easily verify each step of this derivation by replacing Σx in the first line by its equivalent full form, $x_1+x_2+...+x_n$. He will in this manner establish the following rules for himself:

- When summation of an expression involving terms joined by plus or minus signs is called for, the summation may be made by individual terms.

$$\Sigma(a+bx-y)=\Sigma a+\Sigma bx-\Sigma y.$$

- Summation of a constant is equal to N times that constant, if the summation is over N elements.

$$\Sigma a=Na.$$

- Summation of the product of a constant and a variable is equal to the product of the constant and the summation of the variable.

$$\Sigma bx=b\Sigma x$$

Note also:

- Summation of the product of two variable magnitudes cannot be rearranged.

We may now return to the frequency distribution of the protein determinations, reproduced anew in the first two columns. Can we reach a simple scale to replace the awkward x values by coding according to the foregoing scheme? Very simply! First take one of the values of x and use it as the constant a.

Two such selections are used in the table: (1) a_1 is taken as the lowest value of x; (2) a_2 is taken as a value near the center of the series, usually the modal class value. The respective series of residuals, $x-a$, are given as columns (3) and (5). Now take as the constant b the interval on the residual scale, which is of course the true class range on the original scale. Dividing the residuals by b in each series gives the code scales x_1 and x_2 of columns (4) and (6). The contrast of the simplicity of these code scales with the original class centers surely needs no emphasis.

Statistical organization of the data has made possible the expression of magnitude for the variable in the simplest possible form for computational purposes. The new scales, although arbitrarily determined, do not lack descriptive value in the least. For instance, 8 on scale x_1 means 8 steps of 0.5 above the value 10.225, and –3 on scale x_2 means 3 steps of 0.5 below the value 13.225.

The mean value of the series on each new scale may now be secured with easy mental arithmetic. Proceeding as on the sums of variates within classes, fx_1 and fx_2, are given in columns (7) and (8). Dividing the grand totals in each case by the total number of cases (1,937) gives x+1 = 6.016, and x+2 = 0.016.

Table: Illustration of the Principle of Coding by Linear Transformation of Scale for a Frequency Distribution of Uniform Class Range

Class-	Fre-	Coding schemes				Class totals		
centers	quency	(1)		(2)				
		a_1 =	b =	a_2 =	b =			
		10.225	0.5	13.225	0.5			
x	f	$x-a_1$	x_1	$x-a_2$	x_2	fx_1	fx_2	
(1)	(2)	(3)	(4)	(5)	(6)	(7)	(8)	
10.225	4	0.0	0	–3.0	–6	0	–24	
10.725	18	0.5	1	–2.5	–5	18	–90	
11.225	65	1.0	2	–2.0	–4	130	–260	
11.725	157	1.5	3	–1.5	–3	471	–471	
12.225	270	2.0	4	–1.0	–2	1,080	–540	
12.725	327	2.5	5	–0.5	–1	1,635	–327	–1712
13.225	334	3.0	6	0.0	0	2,004	0	
13.725	276	3.5	7	0.5	1	1,932	276	
14.225	212	4.0	8	1.0	2	1,696	424	
14.725	141	4.5	9	1.5	3	1,269	423	

15.225	79	5.0	10	2.0	4	790	316	
15.725	31	5.5	11	2.5	5	341	155	
16.225	14	6.0	12	3.0	6	168	84	
16.725	7	6.5	13	3.5	7	91	49	
17.225	2	7.0	14	4.0	8	28	16	
								+1743
Total	**1,937**					**11,653**		**+31**
Means						6.016		0.016

According to these values the mean is 6.016 steps of 0.5 above 10.225, or 0.016 step of 0.5 above 13.225. In algebraic form,

$x\pm = a + bX\pm\cdot$

That is, $x\pm = 10.225 + 0.5(6.016)$

$= 13.233;$

or again, $x\pm = 13.225 + 0.5(0.016)$

$= 13.233.$

As an exercise the reader may calculate $x\pm$ for the finger-length series given, first using the true scale without coding, then using one or both of the coding schemes just given. This amounts to the constant b being taken as unity, so that $x\pm$ is equal to a plus $X\pm\cdot$ In the original scale is itself of the simple type; any coding, if undertaken, would be confined to taking a equal to 7, say, so that each multiplication would involve a one-digit figure.

The reader will quickly perceive that the columns (3) and (5) are quite unnecessary in routine coding of the x scale in a frequency distribution table, provided that the seriation is uniform. Whichever type of code scale is desired may be written in directly and the values then recorded for the constants a and b involved. Thus, a is the value on the original scale corresponding to zero on the code scale, and b is the true class range (or interval between class centers) on the original scale.

The advantage of taking a near the middle of the series of class centers, giving the smallest values possible on the code scale, is offset to some workers by the resulting introduction of minus signs into the computations. When the number of classes is much in excess of 12, however, a code scale with zero towards the center may facilitate mental arithmetic quite considerably. Some workers using code scales with negative values choose to make a guess at the mean on the original scale, then take *a* as the *x* value immediately below this. If the guess has been good, X + will be a very small but positive value. Should X + prove to be negative in any case, however, it must be remembered that its sign is essential to it in any further calculation.

A seriation involving irregular grouping often makes full coding according to the above scheme of very little if any value. However, the first step alone of subtracting a suitable value *a* may help computations considerably. This abbreviated coding is often used when seriation of data is not worth while and a constant may be subtracted readily by mental arithmetic as each variate magnitude is handled. In such cases, *a* is best taken as a round number somewhat less than the lowest variate so that mental subtractions will give immediately obvious residuals.

3

Frequency Discordance

FREQUENCY MEASUREMENT

One of the great intellectual contributions of Gregor Mendel to biological science was that he foresaw and demonstrated the possibility of stating the results of biological experimentation in simple mathematical expressions. Realizing that forces of chance incidence obscured agreement between the results of his plant hybridization experiments and those he theoretically anticipated, he laboured diligently at repetition of his work to minimize the effect of chance through having large numbers.

Finally he was convinced and felt that others would be convinced of the adequacy of his theory on the basis of his great and precious store of evidence. Yet Mendel did not ultimately secure perfect agreement between the observed frequencies of occurrence of his phenotypes and those which he theoretically expected. The actual differences were relatively small. He attributed them to what the statistician would call "errors of sampling" and rested his hypothesis on the ground that the observational facts approached the theoretical conception more and more closely as the number of observations was increased.

Mendel lacked an objective criterion of the goodness of agreement between observed and theoretically expected frequencies. It was not until 1900, 34 years after Mendel published his work and in the year when it was rediscovered, that Karl Pearson provided a mathematical criterion to meet Mendel's need. Pearson was not concerned specifically with Mendel's problem, for he undoubtedly did not know of it when his memoir was written. He was considering such problems in general, and the discussion immediately following will be given with the same general outlook, illustrated with specific examples.

Considerable attention has already been directed to the fact that, if the probability π of an event is known, and if such events and their alternatives may occur independently in groups of size n, then the probabilities of the possible group combinations are given by the numerical values of the separate terms in the expansion of the binomial $[\pi + (1-\pi)]^n$. Such an expansion, however, yields ideal values to which actual observations can approximate only within the errors of random sampling. If two "perfect" coins be randomly tossed 4 times, the results will not necessarily be 2 heads, head and tail, and 2 tails, in the ratio of 1: 2: 1.

Each toss will yield one of these combinations, but the results of the second, third, and fourth tosses will be quite independent of what happened previously. In the long run, however, as the number of repetitions of the experiment is increased, closer and closer approximation to the theoretical ratio of 1: 2: 1 will be reached. Lack of exact agreement between the observed and theoretical frequencies must be expected purely through chance, and such discrepancies will not in any sense invalidate the theoretical frequencies.

Admitting that exact agreement between observational and theoretical frequencies should not be expected in any finite experience, one is faced with the difficulty of distinguishing between agreement which is satisfactory and that which is unsatisfactory. Any one person's judgment will probably not agree with that of all others.

The theoretical expression may indeed fail to portray the true distribution law. This brings one face to face with a problem of the utmost importance. One must have an objective criterion of the closeness of agreement of the observed with the theoretical frequencies. Without such a test the critical worker may well remain hopelessly confused as to whether the observations fully substantiate the theoretical distribution of frequency or not.

Consider the general case of two series of frequencies distributed into identical classes, the one series arising from observation and the other from theoretical deduction. Within each class let o designate the observed frequency and c the calculated frequency, the total frequency for each series being N. Then $o-c$ will measure the absolute discrepancy in each frequency class. Let n' designate the number of such classes. Then the simplest possible summary of the discrepancies will be secured by adding the n' values of $o-c$. But

$$\sum^{n'}(o-c) = \sum^{n'} o - \sum^{n'} c$$
$$= N - N$$

This summation obviously defeats the purpose; on the whole there will always be no discrepancy by this method. The difficulty again is a matter of signs. It may be surmounted as before by considering the second moments instead of the first. The total discrepancy may be measured as $\sum^{n'}(o-c)^2$, a positive quantity of magnitude E, say. Now E will vary in the aggregate with the magnitude of the individual differences, and hence E would appear to fulfill the requirement of measuring the discrepancy between observation and theory. However, it suffers one severe limitation. The *importance* of any given absolute discrepancy depends on the number expected. If theory should call for a frequency of 500 in a certain group and 502 cases are observed, the concordance is relatively very much greater than if 5 cases were expected and 7 were observed. Since the discrepancies in the n′ classes must finally be summed, the measures of discrepancy should be in comparable terms. The frequency discordance should then be measured on a relative basis, that is, as

$$\frac{(o-c)^2}{c}$$

or

$$\frac{(o-c)^2}{o}$$

In choosing between these two forms of measuring the relative discrepancy in each class, two considerations are of predominant importance. First, when *c* is small in magnitude o may conceivably often be zero solely through errors of sampling. In such cases the relative discrepancy would be indeterminate if,

$$\frac{(o-c)^2}{o}$$

is used. To any suggestion that *c* may be zero and *o* a greater magnitude, thus making,

$$\frac{(o-c)^2}{c}$$

indeterminate, the answer must be clear. If theory does not provide for a frequency that does occur, then the theory is obviously inadequate and it is foolish to attempt to measure concordance. Quite apart from this issue of determinability of the measure itself, there is the second consideration that the calculated frequency c represents a wider experience or broader conception than the frequency o shown by the finite sample; c should then form the logical reference value.

The value,

$$\frac{(o-c)^2}{c}$$

is then proposed as a measure, comparable from one situation to another, of the importance of discrepancies between theoretical and observed frequencies. One may with benefit consider this measure of frequency discordance in relation to a numerical example. If 6 coins were tossed at random 64 times, the number of "heads" being recorded for each toss, a frequency distribution could be prepared for the observed numbers of heads per toss. Now 6 coins, each with a probability of 0.5 of showing the "head" face after random tossing, must show an average distribution of heads in 64 tosses as given in the a column of Table 35. The column o in that table gives frequencies empirically selected as quite typical of the results of an actual tossing experiment. The differences *o–c* vary from zero to 4, the derived elements,

$$\frac{(o-c)^2}{c}$$

varying from zero to a maximum of 1. The magnitudes of the latter elements in relation to the importance of the deviations *o–c* comprise the matter of immediate interest. It will be profitable to study this relationship with care.

Table: Illustrating the Comparability of the Elements $\frac{(o-c)^2}{c}$ as Measures of Frequency Discordance

Head per toss	Frequencies Calculated *c*	 Observed *o*	Discrepancies *o–c*
6	1	0	–1
5	6	8	+2
4	15	12	-3
	20	24	+4
2	15	14	–1
1	6	5	–1
0	1	1	0
Totals	**64**	**64**	**0**

The following two points may well be noted from the data.

1. The three unit deviations (for 6, 2, and 1 "head") yield,

$$\frac{(o-c)^2}{c}$$

values inversely proportional to the expected frequency. That is, as the expected frequency goes up, the importance of any *fixed* deviation from it of observed frequency goes down. This was deliberately chosen as a requirement of the measure of frequency discordance.

2. When the deviation *o–c* is *proportional* to the expected frequency, as in the "4 heads" and "3 heads" groups, the values of,

$$\frac{(o-c)^2}{c}$$

are directly proportional to the expected frequencies. This most important quality of the measure of frequency discordance arises through squaring the deviation but retaining the denominator c in the first power. Just as a deviation of 2 from an expected 500 is of far less consequence than a deviation of 2 from 5, so a deviation of 200 from 500 is of far more consequence than a deviation of 2 from 5.

The measure suggested by Pearson therefore fulfills very simply and neatly the two outstanding logical requirements for a widely comparable measure of discordance between an observed frequency and its theoretical equivalent.

Establishment of an elemental measure of the lack of agreement between the frequencies pertinent to each class permits consideration of the ultimate problem, that of measuring the discordance in all classes as a whole. One wishes to know the probability that the deviations of observed frequencies from the theoretical in the distribution as a whole are attributable to random sampling effects alone. The elements,

$$\frac{(o-c)^2}{c},$$

each being proportional to the general importance of the deviations, may well be totaled for the entire set of classes. That sum forms the criterion designated by Pearson as x^2 (called "chi squared"):

$$x^2 = \sum^{n'} \left[\frac{(o-c)^2}{c} \right]$$

In any practical problem of measuring total frequency discordance, such as that presented, for instance, x^2 is a numerical quantity without dimension of any sort. It is a pure number derived from pure number frequencies. The scale of x^2 is obviously limited at zero as far as smallness is concerned, but may assume very large values when there is great discrepancy between the observed and theoretical frequencies. It is the smallness, numerically speaking, of x^2 which reflects the degree of concordance of observed and calculated frequencies. There is some limit above which values of x^2 will indicate lack of satisfactory agreement between observation and theory. This limit, however, will not be any fixed number applicable to all situations, for x^2 is a sum of basic elements and its magnitude must inevitably be a function of the number of those elements. One may choose to calculate a quantity such as x^2 with the sole objective in view of determining the probability that the total discrepancy it measures might be due to errors of random sampling. It is necessary then to ascertain the random sampling distribution of x^2 derived from n' deviations so that one may know how large its value may become solely through the chance errors of sampling.

This rather complex problem was solved by Karl Pearson as part of his classical memoir establishing the criterion. Tables of the probability integral of x^2 for values of n' from 3 to 30 were then prepared by Elderton, from which it is possible to determine very easily the probability that a given value of x^2 will be exceeded through random sampling errors alone.

In his formulation of the sampling distribution of x^2, Pearson recognized, that to a limited extent, the calculated frequencies are forced to agree with the observed frequencies by the necessary requirement that *c* must equal *o* for comparability. The reader may readily grasp this point by returning, empirically altering therein the values in the c column. He will find that, of the 7 values, he may change only 6 independently at will, the seventh always being determined by the requirement that *c* must equal 64 if the series of inscribed frequencies is to be comparable with the observed one. That is, only 6 of the 7 deviations o-c are independent, for the seventh must be adjusted so that the total, $(o - c)$, equals zero. Of the n' elements entering the sum called x^2, only $n'-1$ are then truly independent.

Pearson's equation and Elderton's derived tables take account of this. It was not until nearly a quarter of a century later that Fisher demonstrated that recognition of other restrictions limiting the number of independent elements should properly and may readily be made. Fisher pointed out that each independent statistic of the observed set of frequencies which is used in formulation of the calculated set imposes one restriction on the freedom of the c values to deviate from the *o* values. The number of independent deviations, n, in the x^2 sum is the total number of elements, n', minus this number of restrictions, r,

$$n = n' - r.$$

The logic of this concept is unquestionably sound, although much acrimonious debate springing from conflicting definitions followed its introduction. Let the reader consider for a moment the distribution of sex at birth. If we assume that the probability of male sex is 0.5, then for Geissler's series we have:

Sex	**Frequencies**		$o-c$	$\frac{(o-c)^2}{c}$
	Observed	**Calculated**		
Males	2,582,914	2,508,816	74,098	2,188.5
Females	2,434,718	2,508,816	-74,098	2,188.5
Totals	**5,027,632**	**5,017,632**	**0**	x^2**=4,217.0**

There is marked discordance between observation and theory, and x^2 is not zero. However, the two elements in the sum are not independent:

$$\frac{(o-c)^2}{c}$$

The *o–c* values are numerically identical and opposite in sign, and this must of necessity be true. There is only one independent deviation, and the magnitude of x^2 must be interpreted on this basis. By referring to appropriate tables, one finds that the probability of the discordance being due to chance is infinitesimal. What then is the true probability of male sex at birth? If one now uses $p=0.514768$ derived from the observed data, then obviously c will agree with o in each case and there will be no discordance. x^2 equals zero. Does this prove that the value $p=0.514768$ is correct? Not at all! Theory has been forced to agree with observation here in two respects, $N_o=N_c$ and $p_o=p_c$, and any independent deviation is no longer possible.

The principle illustrated above in simple terms may be extended to all situations. The number of independent deviations contributing to the x^2 sum is crucial to determination of the probability that the lack of complete concordance between the observed and theoretical frequencies might be due solely to errors of random sampling. This number of "independent deviations" is also designated as the number of "degrees of freedom," an expression that was current in the literature of statistical mechanics several decades ago. Fisher's adoption of the "degrees of freedom" terminology has resulted in its wide dissemination in the statistical literature of the last decade.

This writer suggests that the alternative term "independent deviations" is likely to be more comprehensible to those whose familiarity with mathematical theory is not very extensive. It will therefore be preferred in this discussion.

The sampling distribution of x^2 with n independent deviations is defined by the equation,

$$w = C(x^2)^{\frac{n-2}{2}} e^{-\frac{1}{2}x^2}$$

where C is a constant determined solely by n. This equation, established by Fisher, is identical with the one derived earlier by Pearson, subject to the condition that Pearson's n' be replaced by n+1. Thus Elderton's tables of the probability integral of x^2 still hold, provided that n' therein be read as $n+1$ rather than as the number of frequency classes.

It is of interest to note that the underlying principle of adjusting for imposed restrictions was recognized in other connections at least as early as 1823.

The two shaded tails of the curve indicate by their truncating ordinates the values of x^2 below and above which respectively only 5 per cent of values due to random sampling may be expected to occur when there are six independent deviations.

Thus, values of x^2 below 1.635 (for six independent deviations) are just as unlikely as values above 12.592. One is normally concerned solely with the latter type of value, however. It indicates a point above which values of x^2 have so low a likelihood of occurrence solely through errors of sampling as to cause one to believe that, when they arise in practical work, the theoretical and observed sets of frequency really are not in good enough agreement to substantiate the theory.

Elderton's table gives to seven places of decimals the relative area in the tail beyond (greater than) successive values of x^2 from zero to 70. When more detailed probabilities are desired, Elderton's table may be consulted, it being remembered that therein n' corresponds to our n+1. The mathematically adept reader will discern from equation (3) that the distribution of x^2 is of skew form. Its mean is always at $x^2 = n$, the mode being two units less, that is at $x^2 = n-2$. The curves for each value of n all start at $x^2 = 0$, and extend to infinite magnitudes. For small values of n the skewness is very marked.

However, as n increases, the skewness gradually diminishes and the normal curve is approached as the limiting form of the x^2 distribution. This feature is not readily demonstrable with a diagram drawn on the x^2 scale because the visible range becomes very extended as n increases, with consequent fall in the ordinate values in general. This is demonstrated, wherein only four consecutive values of n are employed. The difficulty may be circumvented by transforming to the scale presents a series of curves so drawn.

$$\frac{x^2}{n}$$

The transition towards normality as n increases may be appreciated more easily from study of this diagram. In practical work, however, n is rarely large enough for the normal curve to provide an adequate approximation to the random sampling distribution x^2.

They serve to illustrate the most important point that the probability of any particular value of x^2 being exceeded is highly dependent on the number of independent deviations embodied in its magnitude.

The practical application of the x^2 test is very simple once the theoretical frequencies are evaluated. For the first two of the following illustrations, the theoretical frequencies are secured from the binomial expansion. In the third and fourth examples the theoretical frequencies are defined by continuous frequency curves. The reader will find another set of data to which he may apply the x^2 criterion as a test of his command of the reasoning involved.

Example: The use of the binomial expansion dealt with an experimental situation in which seeds of Pima cotton were planted 6 to a hill in 1,120 hills. The adjoining calculated values c in that table are those derived by the binomial expansion and are copied directly.

The lack of agreement between the observed frequencies and those given by the binomial expansion may be measured in terms of x^2. Is x^2 sufficiently large to warrant the lack of agreement being designated as "significant," and what does "significant" mean here?

The calculations reveal a very large value of x^2 in this case. Though there are 7 classes in the sum, it contains but 5 independent elements, for the c distribution has been forced to agree with the o distribution with respect to two parameters, N and π.

A glance under $n = 5$ this volume is sufficient to indicate that, with x^2 so large, the probability of the discordance in frequency being due to chance must be essentially infinitesimal.

Study of the o–c values reveals that the seedlings were more abundant in the highest and lowest survival classes, and correspondingly deficient in the intervening classes, than the independence requirement of the binomial permits.

Thus they tended to survive or die as groups rather than as individual plants. Such a situation might arise through physical obstacles like patchy hard soil-crust preventing emergence of the seedlings as groups, or through the presence of diseases tending to take all seedlings in the groups they infect.

The statistical analysis in terms of x^2 merely signifies, of course, that the fates of seedlings within the groups are far from being independent.

Table: The X^2 Criterion Used as a Test of Individual Independence In Survivorship of Grouped Plants X = Number of Surviving Seedlings from 6 Seeds per Hill.

x	Frequencies		Discrepancies $o-c$	$\frac{(o-c)^2}{c}$
	o	c		
0	262	22.4	–239.6	2562.86
1	83	123.4	– 40.4	13.23
2	99	283.8	–184.8	120.33
3	164	348.2	–184.2	97.44
4	217	240.3	–23.3	2.26
5	191	88.4	–102.6	119.08
6	104	13.6	+ 90.4	600.89
Totals	1,120	1120.1	0.0	x^2=3516.09

$n = 5.$

Px^2

is infinitesimally small.

Example. In an effort to demonstrate the applicability of the binomial theorem in forecasting the results of experiments, Weldon tossed 12 dice 26,306 times with thorough randomization between each throw. The appearance of either the 5 or the 6 face was recorded as a favourable event, π thus being established on *a priori* reasoning as one-third. The second and third columns present the observed and calculated frequency distributions of number of favourable events per throw. Are the experimental results in accord with theoretical expectancy?

There are 13 possible combinations of favourable and unfavourable events in this problem. The expected frequencies for 10, 11, and 12 favourable events are, however, quite small. These classes have been grouped because the random sampling distribution of x^2 involves an assumption that the expected frequency shall always be large enough to give a reasonably normal distribution of its sampling error. This is undoubtedly achieved when c is 20 or above, but one will not err far if c is around 10 or more. Grouping the last 3 classes brings c to 14.3, which will be taken as an acceptable value. n' is reduced to 11 by this grouping, and the one restriction of equal total frequencies makes n equal to 10. With x^2 equal to 35.6, P may be found from Elderton's table (entered under $n' = n + 1 = 11$) to be 0.00054. There is therefore only about 1 chance in 2,000 that so large a discrepancy would arise through errors of random sampling alone.

Table: Application of the x^2 Criterion to Weldon's Dice-Tossing Experiment. Theoretical Frequencies Based on the Binomial Expansion Using p = 1/3 x = number of favourable results per toss.

x	Frequencies		Discrepancies o-c	$\frac{(o-c)^2}{c}$
	o	c		
0	185	202.8	– 17.8	1.56
1	1,149	1,216.5	– 67.5	3.75
2	3,265	3,345.4	– 80.4	1.93
3	5,475	5,575.6	– 100.6	1.82
4	6,114	6,272.6	– 158.6	4.01
5	5,194	5,018.0	176.0	6.17
6	3,067	2,927.2	139.8	6.68
7	1,331	1,254.5	76.5	4.67
8	403	392.0	11.0	.30
9	105	87.1	17.9	3.68
10				
11	18	14.3	3.7	.96
12				
Totals	**26,306**	**26,306.0**	**0.0**	**x^2=35.53**
	n = 10.	$P_{x^2} = 0.00054$		

Lack of independence in such a carefully executed random tossing experiment is not likely. One is accordingly disposed to explain the discrepancies between o and c by assuming that the dice were biased. This may be tested by calculating the probability p of the favourable event from the experimental throws.

$$p = \frac{\Sigma ox}{12(26,306)}$$

$$= \frac{106,602}{315,672} = 0.3377$$

A value decidedly in excess of that of one-third given by the assumption of unbiased dice. The statistical significance of the deviation, $p-\pi$, may be tested by applying the technique given in the previous chapter.

For $\pi = 1/3$ and $N = 315,672$,

$$\sigma p = \sqrt{\frac{1}{3} \times \frac{2}{3} \times \frac{1}{315,672}} = 0.000839$$

Then,

$$k = \frac{p - \pi}{\sigma p} = 5.2$$

for which the corresponding P_k is approximatley 1 in 5,000,000. There is no doubt then that Weldon's dice must have been biased.

The technique of tossing may now be tested for its randomization of throws by determining a new theoretical set of frequencies having the same probability of the favourable event as the observed set. Those frequencies, together with the calculations for x^2. There being two restrictions this time, n is equal to 9.

Table: The x^2 Criterion Used as a Test of Individual Independence of Dice in Weldon's Experiment $a = 0.3376986 = p$ observed. x = number of favourable events per toss.

x	Frequencies		Discrepancies $o-c$	$\frac{(o-c)^2}{c}$
	o	c		
0	185	187.4	-2.4	0.03
1	1,149	1,146.5	2.5	0.01
2	3,265	3,215.2	49.8	0.77
3	5,475	5,464.7	10.3	0.02
4	6,114	6,269.4	-155.4	3.85
5	5,194	5,114.7	79.3	1.23

6	3,067	3,042.5	24.5	0.20
7	1,331	1,329.7	1.3	0.00
8	403	423.8	-20.8	1.02
9	105	96.0	9.0	0.84
10				
11	18	16.1	1.9	0.22
12				
Totals	**26,306**	**26,306.0**	**0.0**	**x^2=8.19**

$n = 9.$

$P_{x^2} = 0.52$

For x^2 equal to 8.19, the probability of the frequency discordance being due to chance is seen to be slightly in excess of 0.5, indicating excellent agreement. Thus Weldon's tossing was good. Probably nearly all dice are biased, as material removed in engraving the "dots" is seldom replaced by an equal amount of substance of the same specific gravity.

Example: An experiment to demonstrate the sampling distribution of means has been reported. Therein the theoretical normal curve superimposed on the histogram for the mean of 786 samples of 4 individuals each randomly drawn from a defined supply. The frequencies within specified ranges for both histogram and curve are given. Is the histogram in agreement with the theoretically prescribed curve? Only one of the three parameters defining the normal curve was drawn in this case from the observed frequencies. This restriction exists in every x^2 test, for the total observed and calculated frequencies must agree if the test is to be applied.

Table: Test of Goodness of Fit of the Theoretical Normal Curve to the Distribution of Means from 786 Random Samples of 4.

Class range	Frequencies	Discrepancies o–c	$\frac{(o-c)^2}{c}$
	o	c	
→10.85			

10.85-10.95	16	11.5	4.5	1.76
10.95-11.05	21	15.9	5.1	1.64
10.05-11.15	17	30.5	-13.5	5.98
11.15-11.25	55	51.4	3.6	0.25
11.25-11.35	75	76.0	- 1.0	0.01
11.35-11.45	101	98.6	2.4	0.06
11.45-11.55	123	112.0	11.0	1.08
11.55-11.65	107	111.6	- 4.6	0.19
11.65-11.75	88	97.5	- 9.5	0.93
11.75-11.85	88	74.7	13.3	2.37
11.85-11.95	41	50.2	- 9.2	1.69
11.95-12.05	30	29.8	0.2	0.00
12.05-12.15	14	15.3	- 1.3	0.11
12.15-12.25				
12.25'!	10	11.0	- 1.0	0.09
Totals	**786**	**786.0**	**0.0**	**x^2=16.16**

$n = 13.$

$P_{x^2} = 0.30$

The mean and standard deviation of the curve are those theoretically prescribed; they are not derived from the observed distribution. The derived X^2 of 16-16 therefore has n = 13 independent elements. The theory in therefore well substantiated by the outcome of this sampling experiment.

Example: The 500 determinations by Birge of the width of a spectral band of light have been graduated by a normal curve. The classified frequencies are given in Table below. Does the random error of measurement in determining the width of the light band appear to follow the normal law?

Table: Classified Frequencies

x	**Frequencies**		**Discrepancies** $o-c$	$\frac{(o-c)^2}{c}$
	o	**c**		
→1705				
1705-1725	17	18.3	-1.3	0.09

1725-1745	43	36.1	6.9	1.32
1745-1765	61	70.6	-9.6	1.31
1765-1785	105	101.8	3.2	0.10
1785-1805	103	108.5	-5.5	0.28
1805-1825	89	85.3	3.7	0.16
1825-1845	54	49.5	4.5	0.41
1845-1865	19	21.3	-2.3	0.25
1865-1885				
1885→	9	8.6	0.4	0.02
Totals	**500**	**500.0**	**0.0**	x^2**=3.94**

$n = 6.$

$P_{x^2} = 0.68$

The question to be answered here deals solely with the form of a distribution and is not concerned with the values of the parameters defining the theoretical curve. One must logically use the curve of "best fit," that is, the one whose definitive parameters are derived from properties of the observed data. When fitting a normal curve by moments, this involves setting up the following identities:

Sample Statistic	Curve	Parameter
Frequency	N	$= N$
Mean	x +	= ¼2
Standard deviation	s_x	$= \tilde{A}_x$

Thus the curve used is forced to agree with the observed data in 3 respects, leaving $n = n'-3$ independent elements in the x^2 test. Since n' is 9, then the number of independent deviations left in x^2 is 6. Now the derived x^2 equals 3.94, and one may ascertain IV that the probability of equal or greater departure from normality arising solely through chance is 68 in 100. The theory of normal distribution of error in this case is very well substantiated by the measurements made. The graduation by a normal curve of the observed data on examination scores forming the histogram may be tested as above. The data and calculations of that test are given because of the special interest of the resulting probability, $Px^2 = 0.98$. The graduation given by the normal curve is most unusually good in this case. Only once in 50 times would a better fit arise in a random sample. Is it "too good," thus placing the hypothesis of a normal distribution of the scores in jeopardy?

Table: Test of Goodness of Fit of the Normal Law to the Distribution of Individual Scores on a "Current Affairs" Examination

Score on test	Frequencies		Discrepancies *o–c*	$\frac{(o-c)^2}{c}$	
	o	*c*			
→49		2.9			
50-54	9	6.8	9.5	0.7	0.05
	55-59	18	17.3	0.7	0.03
60-64	36	35.4	0.6	0.01	
65-69	59	58.3	0.7	0.01	
70-74	74	77.3	-3.3	0.14	
75-79	88	82.9	5.1	0.31	
80- 84	72	71.6	0.4	0.00	
85- 89	44	49.7	-5.7	0.65	
90- 94	30.27.9	2.1	0.16		
95- 99					
100-104	19	18.9	0.1	0.00	
105→					
Totals	449	449.0	0.0	$x^2 = 1.36$	
	$n = 7.$ $P_{x^2} = 0.98$				

It is true that this experience deviates from average expectancy of $P_{x2} = 0.50$ just as much as a probability of 0.02, which might well be taken to designate a "significant deviation." Does it also signify that something is wrong? There are some who would argue so, but the burden rests with them to suggest on logical grounds wherein the agreement has been forced. In this specific example there are no grounds for such argument. It must be remembered that such probabilities do arise through chance, just as sometimes those below 0.02 do. One may well "raise the eyebrow" at statistical demonstrations that look too good, but in so doing the skeptic should not neglect the responsibility so assumed,

$$P_{x^2}$$

The point should not be overlooked, however, that, when is extremely high, consideration of the *possibility* that factors forcing concordance may be operative is logically in order. The x^2 test may justify examination of the possibility, but it does not logically signify forced concordance in the same sense that a low value of signifies lack of satisfactory agreement,

$$P_{x^2}$$

If lack of satisfactory agreement were not *a priori* a logical alternative hypothesis to that of good agreement, then there would have been no point in testing the goodness of agreement at all.

4

Variation Measurement

RANGE

It is difficult to understand why statisticians commonly limit their inquiries to Averages, and do not revel in more comprehensive views. Their souls seem as dull to the charm of variety as that of the native of one of our flat English counties, whose retrospect of Switzerland was that, if its mountains could be thrown into its lakes, two nuisances would be got rid of at once. An Average is but a solitary fact, whereas if a single other fact be added to it, an entire Normal Scheme, which nearly corresponds to the observed one, starts potentially into existence.

One would have little ground for castigating statisticians today in the words used by Galton nearly 50 years ago, and yet his eloquent portrayal of the loss sustained by failure to describe variation has imparted lasting quality to that once-merited criticism. To attempt to describe a series of magnitudes solely in terms of some typical value and thus sacrifice all knowledge of their variability is tantamount to obliteration of a most crucial feature of the character being measured. As a basis of generalization in science, each sample statistic such as a mean has its degree of trustworthiness which depends most fundamentally on the amount of variation present in the recorded values from which it is computed.

How enriched an average becomes when a measure of its sampling stability is added to it! The biologist should certainly not fail to measure the variation he encounters in his quantitative observations. Thus may he learn to defend himself against both the embarrassments of overhasty inference and the losses of overwrought caution. Reference to the amount of variation shown by the graph of a frequency distribution would incite contemplation of the "amount of spread" of the distribution as it may be read in terms of the base scale. The *range of variation*, or difference between the minimum and maximum values in any seriation, might well be offered as a first measure of variation. In terms of a histogram this is a finite measure, but if attention is given to frequency curves the appeal of the range fades away like the frequencies at each end of the distribution. In the supply portrayed by the curve, just where does frequency stop? In most cases there is nothing harder to determine with real satisfaction than the answer to this question.

The range of any finite sample is of course quite definite. But what is one to do about the outlying case which appears so commonly to plague the investigator with conjecture as to its "really belonging" with the rest of the individuals in the sample? Exclusion of that one measurement may change the range markedly. The importance which the extreme individuals assume in this portrayal of variation is overwhelming. Is it of no consequence at all where the intermediate variates fall? One faces many such disturbing problems in careful analysis of the descriptive quality of range as a measure of variation. The need for a fundamentally sounder approach to the general problem soon becomes manifest.

INDIVIDUAL DIFFERENCES

Variations are observed through recognition of differences between individuals. Description of variation may on occasion be undertaken in that way. Variation in the characteristics of the members of a family may well be, indeed generally is, measured by direct comparison of each individual with every other. The justification of this method is due chiefly to two factors: first, a personal interest existing in each member as an entity; and second, most families are of distinctly limited number and the comparisons do not become onerous. The existence of such factors as these represents a very special situation when one turns to study of variation systems in general. Algebraic analysis readily shows that among N individuals the number of,

$$\frac{N}{2}(N-1)$$

different comparisons possible is equal to. When there are but 20 variates, 190 individual differences appear, and 100 variates would yield 4,950 comparisons. The difference between the minimum and maximum variates—the range—is of course just one of the set in each case. To attempt summary of individual differences by considering all possible pairs is obviously out of the question.

It may well have occurred already to the reader that the set of N–1 deviations of one selected variate from all others in any series must give all the information incorporated in,

$$\frac{N}{2}(N-1)$$

the much larger set of individual differences. The scientist is not interested in the latter values as such, but in the amount of variation characterizing the N variates. But if the deviations from some one variate are taken, which one would be most appropriately chosen as this reference standard? This is a troublesome question. No one variate is clearly the best. Indeed, are not all equal in importance?

Need it be one of the N individuals of the series? What could provide a better standard than a value representative of all; for instance. What would be more logical than that, having established a typical magnitude, one should turn attention to variation in terms of deviations of the individuals from that magnitude?

Consideration of the suitability of the three typical values for the purpose quickly eliminates the modal value, leaving the issue of choosing between the median and the mean. Of these, only the mean is fully representative of all variate magnitudes, for the median is just central with respect to total frequency. One may thus be guided to the mean as the standard of reference.

DEVIATION FROM TYPE

Just as the variates, x_1, x_2, x_3..., x_N, formed the basis for the measurement of type, so the quantities,

$$(x_1 - x +), (x_2 - x +), (x_3 - x +)\ldots,(x_N - x +)$$

provide a starting point for the measurement of variation or deviation from type. For brevity, one may designate the deviation of each variate from the *mean* of its series as a *deviate.* The task is to summarize these deviates into a single value that has, as far as possible, the qualities stipulated in the previous chapter for a good descriptive statistic. If these conditions are adequately fulfilled it may serve as our description of the amount of variation in any series.

There is no simpler form of summary than that provided by determining the average value. Would the average of the deviates conform with requirements for the summary of variation? Unfortunately the answer must be in the negative. Since the mean is the "balancing point" of the variates, the sum of the positive deviations from the mean must equal that of the negative deviates, regardless of the amount of variation present in the series. Algebraically,

$$\frac{\Sigma(x - \bar{x})}{N} = \frac{\Sigma x}{N} - \frac{\Sigma \bar{x}}{N} = \bar{x} - \bar{x} = 0$$

This difficulty of positive and negative deviations which just balance one another in the sum may be removed in either of two simple ways.

First, the deviations may be considered without regard to their signs; all deviates may arbitrarily be considered as positive. Second, the deviations may be squared, or indeed raised to any even power, to give quantities of positive sign throughout.

The average of the deviates taken without regard to their signs has been used somewhat as a measure of variation. It is often called the *average deviation*. This quantity is very easy of comprehension and would appear to summarize the deviations satisfactorily. However, it is not widely useful as a measure of variation in more advanced statistical work. This is due to the great difficulty in solving algebraic equations involving quantities whose signs are to be ignored.

Knowledge concerning such procedures is comparatively scarce, and the statistical worker would seriously limit his mathematical field, cutting himself adrift at once from a vast amount of that very logic which it is his desire to turn to his advantage, if he arbitrarily ruled all deviates as positive quantities. Consideration of the positive and negative deviations in two separated groups would also be unsatisfactory, since such a procedure would entail the use of two magnitudes to express the one idea of amount of variation, and would involve any further reasoning based on that idea with considerable complexity.

It is a derived property of numbers that the product of two negative quantities must itself be a positive number. All have learned the elementary rule of algebra flowing from this, that the *square* of any number is always positive, regardless of the sign of that number. The expedient of ridding the problem of the measurement of variation from a vexatious difficulty imposed by the signs of the deviates, through using the squared values rather than the absolute deviations, is very appealing in its mathematical simplicity.

Very early in the quantitative measurement of variation this expedient was resorted to, and the results have proved so useful that the general statistical practice today is to measure variation as a mathematical function of the squared deviates, $(x–x +)^2$. Summing these quantities and dividing by N, one will secure an average value typifying the squared deviations. Symbolizing this average by m_2, for it is commonly known as the *second-moment coefficient*, the equation for the mean squared deviate becomes,

$$m_2 = \frac{\Sigma(x - \bar{x})}{N}$$

In the complete absence of variation, m_2 must obviously be zero. When only a little variation is present, the values of $(x–x +)^2$ are individually small and therefore their average is small. As the values of x differ increasingly among themselves, m_2 must rise rapidly in magnitude. Also, the mean squared deviate is fully representative of all members of the series in that it takes into account every variate in the same manner.

As a numerical quantity this mean squared deviate seems to fulfill those conditions which are logically to be imposed for the measurement of deviation from type. However, variation is in our everyday thinking a matter of simple differences, not of squared differences. Squaring was resorted to for the purely mathematical reason of eliminating signs. The immediate objective being accomplished, one may now reverse the move by extracting the square root of m_2 so as to return to the original scale of measurement.

These adjustments of squaring the deviates, then taking the square root of their mean, comprise transformations to and from a scale of squares made for convenience in hurdling an obstacle.

The resultant quantity is a magnitude on the original scale of measurement and one directly proportional to the amount of variation present. It was called the standard deviation by Karl Pearson in 1894, and indeed it has since then become in fact the standard measure of variation.

A statistic of tremendous utility, the standard deviation lends itself admirably to use in advanced statistical work. Every effort given immediately to comprehend its nature and meaning as a measure of variation will be well repaid. Symbolizing it by s, the fundamental definitive equation may be written in the form

$$s_2 = \sqrt{\frac{\Sigma(x-\overline{x})^2}{N}}$$

CALCULATION PROCEDURE FOR THE STANDARD DEVIATION

The Transfer from Deviates to Variates

The basic definition of the standard deviation is shown at a glance by equation above. This formula is ill suited for computational purposes, however. The deviations from the mean when numerically expressed are generally cumbersome, and squaring them is an onerous procedure providing considerable opportunity for mistakes.

Harris pointed out in 1910 that a simple transfer from

deviates to variates very greatly reduces the work involved. The rearrangement may be shown algebraically as follows:

$$s_x^2 = \frac{\Sigma(x-\bar{x})^2}{N}$$

$$= \frac{\Sigma(x^2 - 2x\bar{x} + \bar{x}^2)}{N}$$

$$= \frac{\Sigma x^2}{N} - \frac{\Sigma 2x\bar{x}}{N} + \frac{\Sigma \bar{x}^2}{N}$$

$$= \frac{\Sigma x^2}{N} - 2\bar{x}\frac{\Sigma x}{N} + \bar{x}^2$$

But,

$$\frac{\Sigma x}{N} = \bar{x}$$

Therefore,

$$s_x^2 = \frac{\Sigma x^2}{N} = \bar{x}^2$$

and

$$s_x = \sqrt{\frac{\Sigma x^2}{N} = \bar{x}^2}$$

Thus the standard deviation of a series of magnitudes may be derived expeditiously from the means of the first and second powers of the variates; the individual deviates need not be computed.

When a calculating machine is available and N is not too large (not in excess of 100, say), Σx and Σx^2 may be accumulated directly on the machine without previous seriation of the data. The mean and mean square may then be determined, the square root of the difference between them giving the standard deviation. Applying this procedure to the erythrocyte counts given by Haden for 40 normal men, the following record sheet of computations was secured.

Series: Haden

x = red blood cell count

$N = 40$

$\Sigma x = 198.91\ x + = 4.97275$

$$\frac{\sum x^2}{n} 24.838283$$

$\Sigma x^2 = 993.5313$

$s_x^2 = 0.110040$

$s_x = 0.331723$

General Summations from Frequency Tables: It seems desirable to emphasize at this point that, in our algebraic notation, Σ has been defined as indicating the sum of *all* the values directly following it. The symbol for the variable, for instance x, when placed after a summation sign is understood to indicate the individual values of x, the sum being made for all variate magnitudes.

The class centers in a frequency distribution really correspond to the variate magnitudes under an assumption that measurement was remade to the nearest value according to the scale of graduations given by those class centers. In the tables to be given immediately following in this chapter, these centers will be designated as x_c solely for purposes of distinction. This permits more convenient demonstration of the point that general summations made from frequency tables proceed by two steps. In securing Σx to calculate the mean, for instance, one first multiplies the class center by the frequency to get the sum of all x values in the class. Then one totals these products. Forgetting that multiplication is simply repeated addition of the same quantity, the reader might easily overlook the preliminary multiplication as being part of the general summation. If there are k classes in a frequency table, then £x as applied to it is identical with,

$$\sum^{k} fx_c \cdot$$

We shall see that not only the first and second but also higher powers of x enter statistical computations, and summation of them called for by the general algebraic notation £x^p reduces to,

$$\sum^{k} fx_c^p$$

in special frequency-table notation. The general rule may be stated as follows: If data are seriated into a frequency table and a sum over all *N* values for the variable is called for, each class center must be counted as many times as is indicated by the frequency of the class.

The Uncoded Frequency Table

The last column, headed fx_c^2 ,is assembled by securing the products of the corresponding values in the x_c and Σx_c columns, its sum being Σx^2 of the general algebraic notation. With a calculating machine there would not be any need to include this last column in the table as the products could be accumulated in a sum directly on the machine as each is secured.

Code Scales

Uniform grouping in a frequency table permits one to use code scales to speed up calculations. The standard deviation may be secured along with the mean on an x scale, the results then being transformed to the original scale. This transformation equation for converting s_x to s_X may be derived very simply as follows: Since $x = a + bx$, and $x +^{-} = a + bX +^{-}$, then $x – x + = b\ (x – X + ^{-})$.

Table: Calculation of Mean and Standard Deviation for Weight (in pounds) at Birth of 402 Infants

Class range	x_c	f	fX_c	fx_c^2
5.0-5.49	5.25	7	36.75	192.9375
5.5-5.99	5.75	28	161.00	925.7500
6.0-6.49	6.25	56	350.00	2,187.5000
6.5-6.99	6.75	72	486.00	3,280.5000
7.0-7.49	7.25	86	623.50	4,520.3750
7.5-7.99	7.75	65	53.75	3,904.0625
8.0-8.49	8.25	42	346.50	2,858.6250
8.5-8.99	8.75	22	192.50	1,684.3750
9.0-9.49	9.25	13	120.25	1,112.3125

9.5-9.99	9.75	7	68.25	665.4375
10.0-10.49	10.25	3	30.75	315.1875
10.5-10.99	1.75	1	10.75	115.5625
Total		**402**	**2930.00**	**21,762.6250**

$N = 402$

$\Sigma x = 2930.00.$ $x + = 7.288557.$

$\Sigma = 21762.6250.$ $\frac{fx^2}{N} = 54.125883$

$s_x = 1.012820.$ $s_x = 1.006390.$

Now,

$$s_x^2 = \frac{\Sigma(x - \bar{x})^2}{N}$$

$$= \frac{\Sigma[b(x - \bar{x})]^2}{N}$$

$$= b^2 \frac{\Sigma(x - \bar{x})^2}{N}$$

$$= b^2 s_x^2$$

Therefore, $s_x = bs_x$

The coding constant a does not enter this transformation equation. Subtraction of a constant quantity from a set of variates will leave the residuals with the same variability as characterized the original measurements.

Table: Calculations as for Table given above but Using Coding

x_c	f	x_c	fx_c	fx_c^2
5.25	7	0	0	0
5.75	28	1	28	28
6.25	56	2	112	224
6.75	72	3	216	648
7.25	86	4	344	1,376

7.75	65	5	325	1,625
8.25	42	6	262	1,512
8.75	22	7	154	1,078
9.25	13	8	104	832
9.75	7	9	63	567
10.25	3	10	30	300
10.75	1	11	11	121
Total	**402**		**1,639**	**8,311**

$N = 402.$ $a = 5.25$ $b = 0.5.$

$\Sigma x = 1639.$ $X + = 4.077114.$

$\Sigma = 8311.$ $\frac{\sum x^2}{N} = 20.674129$ $x + = a + bX +$

$= 7.288557.$

$s_x^2 = 4.051270$ $s_x = 2.012777.$ $s_x = bs_x = 1.006389.$

The simplification of the calculations by transformation of scale may again be attested by the ease with which the work may be verified by mental arithmetic. The final results are of course identical (within last-place modification errors) with those secured.

LEAST SQUARES

In accepting the mean as the typical value about which to measure variation, and then establishing the standard deviation as an appropriate measure of the variation, we have associated two descriptive statistics that are held closely together by a natural mathematical bond which we are now in a position to reveal.

Let us consider the change in the "mean square deviation" about any point a as the latter moves along the scale of measurement.

Instead of subtracting x + from every value of x, let us subtract some other constant, a.

Then the mean square deviation about a will be,

$$\text{M.S.D.} \quad = \frac{\Sigma(x-a)^2}{N}$$

$$= \frac{\Sigma x^2}{N} - 2a\bar{x} + a^2$$

$$= \frac{\Sigma x^2}{N} - \bar{x}^2 + \bar{x}^2 - 2a\bar{x} + a^2$$

$$= s_x^2 + (\bar{x} - a)^2$$

It is immaterial whether a is greater or less than x + as far as the sign of (x +–a)^2 is concerned; this sign will always be positive. Therefore the mean square deviation will always be greater than the squared standard deviation unless a is made equal to x +. From this it follows directly that the mean of any series is the value from which the variates as a whole deviate the least, provided that variation is measured by squaring the deviations.

The standard deviation measures the variation which has been so minimized. Mathematicians express this property of the mean by saying that it is the "least squares solution" of finding a value most closely approximating any series of magnitudes. This affinity of mean and standard deviation may naturally raise the question: If the deviations had been taken without regard to sign instead of squaring them, would the mean still remain as the point about which variation so measured would be a minimum? Let the mean deviation be defined in general form as follows:

$$= \frac{\Sigma|x-a|}{N}$$

wherein the vertical rules indicate that the sign of the quantity enclosed by them is to be ignored. For what value of a is this mean deviation a minimum? The complexity of the derivation necessitates that this question be answered here without proof. The answer is quite certain, however, that the variates deviate least from their *median* value if first-power deviations without signs are used as the measure of variation. It seems very reasonable then that the "average deviation" should be taken about the median rather than about the mean; the median and this mean deviation have the same affinity as the mean and standard deviation.

In view of these facts and the probably greater simplicity of comprehension of the median and mean deviation, it is very reasonable to ask why statisticians work so generally in terms of the mean and standard deviation. The explanation arises from precisely the same type of reasoning as led to the question. We are in search of simplicity. "Least squares" solutions, or the minimizing of squared deviations to reach representative values, may be made with facility, whereas the minimizing of absolute deviations becomes so involved that the problems can rarely be solved that way.

Now problems of defining a moving representative value prove multitudinous in statistical analysis when one passes to the mutual relationships of two or more variables, and therefore consideration of the simplicity with which solution of such problems *in general* may be reached assumes paramount importance. If there is no difficulty in accepting the mean as a representative value, then its logical associate in measuring variation should likewise be acceptable. The use of the standard deviation as a measure of variation may at first offer the student some tasks in comprehension, but it will pave the way to progress where the average deviation would lead into a blind alley.

INDEPENDENT DEVIATIONS AND VARIANCE

During our preliminary discussion of the variation shown by N variates, attention was drawn to the point that fundamentally only N–1 independent differences exist between variates. Of the,

$$\frac{N}{2}(N-1)$$

possible individual differences, N–1 arise through comparing a selected variate with the others. The remainder arise from comparing the others with one another, all these secondary differences being derivable from what we may call the N–1 primary deviations. Take a simple series of 3 variates, x_1, x_2, and x_3. The primary differences may be given as $x_2 - x_1$, and $x_3 - x_1$.

The only possible secondary difference here is $x_2 - x_3$, which is just $[(x_2 - x_1) - (x_3 - x_1)]$, or the difference between the primary differences. This may be extended to a series of any size. No information about variation is gained by considering the secondary differences.

Then deviations are taken from the mean instead of from some selected variate, quite obviously N differences (in this particular case we have called them deviates) arise. Has one additional element of information about variation been gained? Certainly not! Each deviate is to some extent smaller than it would be if we demanded *independent* deviations. Each deviate is, to a small degree, a comparison of a variate with itself, for x + is the sum of the Nth parts of all variates.

The extra deviate simply makes up for this. Fundamentally there are only N–1 independent deviations. One might therefore argue that the sum of their squares should be divided by N–1 instead of by N. This slight change in the divisor may be considered as inconsequential *unless N is small* and one is desirous of estimating the variation in the supply directly from the sample.

The adjustment plays a most important role in the "smallsample" techniques, to the development of which Fisher and others have contributed so notably.

Detailed consideration of these techniques seems to the writer to belong as a sequel to an introductory discussion of statistical reasoning, and development of them will be reserved for another volume.

The reader, however, is now in a position to understand a variant of the formula just given for the standard deviation. The square root of v should *not* be called the standard deviation of the sample, but rather the maximum likelihood estimate (M.L.E.) of the standard deviation $\tilde{A}$, of the supply. It may be derived very simply from s_x by means of a correction factor.

$$\text{M.L.E.}_{\sigma} = S_x \sqrt{\frac{N}{N-1}}$$

The reader may derive this factor for himself very readily from equations.

THE QUARTILES

Description of frequency distributions in terms of their "probability points" has been initiated in the discussion of the median as a central value. Its representativeness depends on its "middlemost" character. There is a very simple extension of this idea that permits of the portrayal of variation in the same terms.

Half of the variates in any series deviate in the positive sense from the median value, the other half being negative deviations. One may choose to establish the "middlemost deviation" in each direction. Such values taken with the median itself clearly divide the whole ranked series into 4 equal frequency groups. They are therefore called the *quartile* values, the lower being known as the first quartile (Q_1), the median forming the second, and the upper value the third quartile (Q_3). The value given by one-half of the difference between the first and third quartiles is used on occasion as a description of the variation present; it is usually called the *semi-interquartile range* and is symbolized as Q. Then

$$Q = \frac{Q_3 - Q_1}{2}$$

It is worthy of note that, like the median, Q is not affected by extreme variates. It is a measure of mediocrity in variation, and is based on frequency of variates rather than individual deviation magnitudes.

When working from a frequency table rather than ranked variates, determination of the quartile values usually involves some calculation because the points must be assigned within class ranges. To clarify these matters an illustration of procedure, utilizing the data on examination scores graphically portrayed. The assumption underlying selection of an appropriate point within a class range is that the frequency of the class is uniformly distributed over the range. Attractive because of its simplicity, this assumption may be replaced, if desired, by others of more elaborate nature giving better approximation to frequency-curve form. We shall use the simple one herein.

A "cumulative frequency" column is formed in which the frequency up to the boundary line between classes is given. The class range within which each quartile must fall then becomes apparent by inspection. That proportion of the scale covered by the class is taken which, under the assumption, yields the necessary frequency. Sixty-three units exist below a score of 65, and 122 below a score of 70. To 65 must be added the proportion,

$$\frac{112.25 - 63}{122 - 63}$$

of the class range immediately above 65. The evaluation of all quartile values is appended to the table. The quartiles plotted on the histogram, each shaded area comprising one-quarter of the whole, the dividing ordinates rising at the quartile values.

Table: Cumulative Frequency and the Calculation of Quartiles (Scores of 449 students in an examination)

Class range	Frequency f	Cumulative frequency Σf
45-49.9	2	2
50.-54.9	7	9
55-59.9	18	27
60-64.9	36	63
65-69.9	59	122
70-74.9	74	196
75-79.9	88	284
80-84.9	72	356
85-89.9	44	400
90-94.9	30	430

95-99.9	11	441
100-104.9	5	446
105-109.9	3	449
$\frac{N}{4} = 112.25$	$Q_1 = 65 + \frac{49.25}{59} \times 5 = 65 + 4.2 = 69.2$	
$\frac{N}{2} = 224.5$	$Q_2 = 75 + \frac{28.5}{88} \times 5 = 75 + 1.6 = 76.6$	
$\frac{3N}{4} = 336.75$	$Q_3 = 80 + \frac{52.75}{72} \times 5 = 80 + 3.7 = 83.7$	

This same principle may be extended to the fractionation of the total frequency into any systematic scheme of parts desired.

Among these, deciles and percentiles are not infrequently used in discussions of educational data. The reader may readily gain some experience with these finer divisions and the problems they present by first applying the above principles of procedure to the data.

RELATIVE VARIATION

For the comparison of the variabilities existent in different series, it is sometimes necessary to express variation on a relative scale. For example, the standard deviation of finger length (left middle finger) for the series is 0.5479 mm. For the same group of individuals, the standard deviations for head length and stature were found to be 6.046 mm. and 2.5410 inches, respectively.

Direct comparison of these standard deviations is not very informative because of the difference in general level of magnitude of these dimensions. Moreover, one is expressed in inches and the others in millimeters.

If, however, the standard deviations are expressed as percentages of the respective means, yielding 4.74 per cent, 3.15 per cent, and 3.88 per cent, respectively, it is seen immediately that finger length is relatively the most variable, and head length relatively the least variable, of the three. This relative measure of deviation is known as the *coefficient of variation*:

$$\text{C.V.} = 100\frac{s_x}{\bar{x}}$$

A word of warning is warranted with reference to the use of this measure. In its interpretation, caution must always be exercised not to neglect the fact that its magnitude is just as much a function of the mean as it is of the standard deviation. Variabilities measured on fully comparable scales should be compared directly in terms of the standard deviations as far as is practical. If a relative measure is necessary in order to circumvent the use of different scales or to allow for incompatibility of means, care should be exercised to assure oneself that it is wholly logical to consider the variation in proportion to the mean value as a basis of comparison.

The use of the coefficient of variation may give valuable information on questions such as the following: Is human stature more variable in the adult than in the new-born infant? Is the error of the chemical determination of ash in foods relatively greater than that for protein in the same materials? It would be quite foolish to compare the error of a measurement process with the variability of individuals so measured, by means of the coefficient of variation.

Type of Variation

The foregoing discussion of variation has been strictly confined to the question of *amount* of variation. Questions of skewness, kurtosis, *et cetera,* have been intentionally divorced from the discussion, for they are concerned with *type* of variation. In considerations of the amount of variation, the signs of the differences between variates are irrelevant; one is concerned solely with consistency of measurements, like the scattering of bullet holes about their focal point on a target. We shall now pass to a general scheme of measuring the characteristic features of frequency distributions in which x + and s_x form the first elements, measures of skewness and kurtosis following directly in the scheme.

5

Measurement Scale

VARIABLE CLASSIFICATION

Stevens developed the measurement scale hierachy into four categories, namely, nominal, ordinal, interval and ratio scales. Stevens further prescribed statistical analyses that are appropriate and/or inappropriate for data that are classified according to one of the four scales above. The nominal scale is the lowest while the ratio scale variables are the highest.

However, this scale typology has seen a lot of critisisms from, namely, Lord, Guttman, Tukey and Velleman and Wilkinson. Most of the critisisms tend to focus on the prescription of scale types to justify statistical methods. Consequently, Velleman and Wilkinson give examples of situations where Steven's categorization failed and where statistical procedures can often not be classified by Steven's measurement theory.

Alternative scale taxonomies have therefore been suggested. One of such was presented in Mosteller and Tukey. The hierachy under their classification consists of grades, ranks, counted fractions, counts, amounts, and balances.

A categorical variable is one for which the measurement scale consists of a set of categories that is non-numerical. There are two kinds of categorical variables: nominal and ordinal variables. The first kind, nominal variables, have a set of unordered mutually exclusive categories, which according to Velleman and Wilkinson "may not even require the assignment of numerical values, but only of unique identifiers (numerals, letters, colour)". This kind classifies individuals or objects into variables such as gender (male or female), marital status (married, single, widowed, divorced), and party affiliation (Republican, Democrat, Independent).

Other variables of this kind are race, religious affiliation, etc. The number of occurrences in each category is referred to as the frequency count for that category. Nominal variables are invariant under any transformations that preserves the relationship between subjects (objects or individuals) and their identifiers provided we do not combine categories under such transformations. For nominal variables, the statistical analysis remains invariant under permutation of categories of the variables.

When nominal variables such as variable gender have only two categories, we say that such variables are dichotomous or binary. These kinds of variables have been sometimes referred to as the categorical-dichotomous. A nominal variable such as colour of hair with categories {black, red, blonde, brown, gray}, which has multiple categories but again without ordering, is referred to as categoricalnominal.

The second kind of variables, ordinal variables, are where the categories are ordered. Using the definition in Velleman and Wilkinson, if S is an ordinal scale that assigns real numbers in $5R$ to the elements of a set P, of observed observations, then

Such a scale S preserves the one-to-one relation between numerical order values under some transformation. The set of transformations that preserves the ordinarily of the mapping has been described by Stevens as being *permissible*. That is, the monotonic transformation I is permissible if and only if:

Thus for this scale of measurements, permissible transformations are logs or square roots (nonnegative), linear, addition or multiplication by a positive constant.

Under an ordinal scale, therefore, the subjects or objects are ranked in terms of degree to which they posses a characteristic of interest. Examples are: rank in graduating class (90th percentile, 80th percentile, 60th percentile); social status (upper, middle, lower); type of degree; etc. The Likert variable such as the attitudinal response variable with four levels (strongly disapprove, disapprove, approve, strongly approve) can sometimes be considered as a partially ordinal variable (a variable with most levels ordered with one or more unordered levels) by the inclusion of the response level "don't know" or "no answer" category. Ordinal variables generally indicate that some subjects are better than others but then, we can not say by how much better, because the intervals between categories are not equal. Consider again the 120 students in our biostatistics class now classified by their status at the college (freshman, sophomore, junior, or senior). Thus the status variable has four categories in this case.

The results indicate that more than 75% of the students are either sophomores or juniors. There is an intrinsic ordering of the categories of the variable "status", with the senior category being the highest level in this case. Thus this variable can be considered ordinal.

On the other hand, a *metric variable* has all the characteristics of nominal and ordinal variables, but in addition, it is based on equal intervals. Height, weight, scores on a test, age, and temperatures are examples of interval variables. With metric variables, we not only can say that one observation is greater than another but by how much. A fuller classification of variable types is provided by Stevens.

Within this grouping, that is, *nominal, ordinal,* and *metric,* the metric or interval variables are highest, followed by ordinal variables. The nominal variables are lowest. Statistical methods applied to nominal variables will also be applicable to either ordinal or interval variables but not vice versa. Similarly, statistical methods applicable to ordinal variables can also be applied to interval variables, again but not vice versa, nor can it be applied to nominal variables since these are lower in order on the measurement scale.

When subjects or objects are classified simultaneously by two or more attributes, the result of such a cross-classification can be conveniently arranged as a table of counts known as a *contingency table.* The pattern of association between the classificatory variables may be measured by computing certain *measures of association* or by fitting of *log-linear model, logit, association,* or other models.

CATEGORICAL DATA EXAMPLES

Suppose the students in this class have been cross-classified by two variables, gender and status, then the resulting table will be described as a two-way table or a two-way contingency table below.

In general, our data will consist of counts {*nπ*, *i* = 1, 2, ..., *k*} in the *k* cells (or categories) of a contingency table. For instance, these might be observations for the *k* levels of a single categorical variable, or for *k–IJ* cells of a two-way 7 × *J* contingency table. We shall treat counts as random variables. Each observed count, *n*, has a distribution that is concentrated on the non-negative integers, with expected values denoted by *rrij* = *E(n, i)*. The {*mi*} are called *expected* frequencies, while the {*rii*} are referred to as the observed frequencies. Observed frequencies refer to how many objects or subjects are observed in each category of the variable(s).

Analyses

For the data, interest for this type of data usually centers on whether the observed frequencies follow a specified distribution usually leading to what is often referred to as 'goodness-of-fit test'. For the data, on the other hand, our interest in this case is often concerned with independence, that is, whether the students' status is independent of gender. This is referred to as the test of independence. An alternative form of this test is the test of "homogeneity", which postulates, for instance, that the proportion of females (or males) is the same in each of the four categories of "status". We will see that both the independence and homogeneity tests lead asymptotically to the same result. We may on the other hand wish to exploit the ordinal nature of the status variable (since there is an hierachy with regards to the categories of this variable).

Example of a 2×2 Table

Suppose the students in the biostatistics class were asked whether they have ever smoked; this would lead to a response variable that we will call here "smoking status" with categories Yes and No. The two-way cross classification of the students by gender and smoking status, leading to a 2×2 contingency table.

The resulting table indicates that of the 120 students in the class, 70 indicated that they have once smoked. The 2×2 table has perhaps received more attention than many other tables. The reasons for this are related to "how the observed frequencies in the table are obtained." In this case, there are only 120 students in this class, and we may therefore consider that this value or sample size is fixed. With 120 fixed, the students are then cross-classified by the two variables, response and gender. This schemme leads to what has been described as the "multinomial sampling scheme." Other sampling schemes relating to the 2×2 table will be examined. Again, interests center here on whether smoking status is independent of gender or whether the response YES is uniform across gender.

Three-Way Tables

Assuming the students were also asked their attitudes towards abortion, this would lead to yet another variable, designated here as attitudes, with three response categories {positive, mixed, negative} designated here as 1, 2, and 3, respectively. We now cross-classified the resulting data as a three-way contingency table having variables gender, status, and attitude, resulting in a 2×4×3 three-way contingency tables having 24 cells. The status categories 1, 2, 3, and 4 refer respectively to freshman, sophomore, junior, and senior. The form of analysis here can be viewed in several ways.

(a) We may wish to consider whether the attitudes towards abortion are independent of gender. In this case, we only need to collapse over the status variable (in other words we will ignore the status variable) and use the methods mentioned in the previous section to analyse the data; that is, we will conduct a test of independence.

(b) We may also be interested in the relationship between the status category and attitudes towards abortion. This again would lead to a 4 × 3 two-way contingency table. Again the usual test of independence will answer this question. Note here that we are collapsing over the gender variable.

(c) However, we may view collapsibility over either the status or gender variable as unsatisfactory because we do not know whether the strength and direction of the association between the resulting two variables will change when collapsed over the third variable (this is called Simpson's paradox).

It is therefore desirable to analyse the data like the familiar analysis of variance for the continous type data where we can study the interactions between the various variables. The equivalent analysis of variance (ANOVA) model in categorical data analysis is the log-linear models analysis, which we shall apply not only to three-way tables but to higher dimensional tables in this book. The procedure allows us to study interactions between the explanatory variables and the response variable. If, however, we consider variable attitude as a response variabe, an even better procedure would be to consider fitting *logit* models to the data. We can further exploit the intrinsic ordinality of the response variable {+ve, mixed, -ve} by fitting specialized models, such as the proportional odds, continution ratio, adjacent category, and cumulative logit models to this data. If we consider again the 4x3 table formed from the three-way table when collapased over gender; we would notice here that both variables "status" and "attitudes" are ordinal in nature. We can study the associations between these two variables better by employing either the row, column or R+C association models. These classes of models, which first appeared in Goodman, models the local odds-ratios in the table.

Consider again, the two-way table formed from collapsing over the "status" variable, that is, where only the response variable "attitude" can be considered ordinal, gender variable being nominal in this case. Thus, we can fit the row association (R) model to this data. Another possible model is the multiplicative RC association model. We consider the analysis of doubly classified data.

For the students in our class we asked the following questions:

- What is your political party affiliation?
- Which party did you vote for in the last Presidential election?

We gave them three party choices only: Democrat, Republican, Independent. We cannot employ the usual two-way table model of independence as presented earlier with these data. We would expect the analysis to be different in this case because it is obvious that the diagonal cells are highly correlated: Democrats tend to vote Democrat and Republicans likewise.

Hence, the models that we would consider must take into consideration the diagonal structure of this table. We shall develop diagonal parameter and similar models necessary for a proper explanation of such data. Other possible models we would want to examine are whether there is symmetry with regards to the cells or symmetry with regards to the marginal totals. Models discussed in this chapter will examine symmetry and marginal symmetry models that are appropriate for this class of data.

A unified approach to fitting these models will be presented. We shall further extend our discussion to methods of analysis for agreement data arising from two and three raters. Nonstandard loglinear techniques and will be discussed for implementing these class of models. In the last series of questions, the students were asked regarding abortion: Should a pregnant woman be able to obtain a legal abortion if:

1. She is married and does not want more children?
2. The family has very low income and cannot afford any more children?
3. She is not married and does not want to marry the man?

The form of these questions was tailored to the 1976 US General Social Survey. Since there are three questions whose responses are simply Yes or No, there are 2^3=8 possible combinations of responses. In this case, there are 27 students in this category. The data fall into the category of repeated binary response data. Because the responses are repeated, there is therefore an intrinsic correlation among the three responses and models that will be employed for this would have to exploit this correlation structure within the subjects' responses. We discuss in the last chapter of this book logit, marginal, and other similar models for handling data of this nature. The discussion in this chapter also covers the case involving more than one covariates with repeated observations coming from clusters of subjects.

Probability Models

Categorical data analysis depends on a thorough understanding of the various sampling schemes that may give rise to either a one-way, a two-way, or a multidimensional contingency tables under consideration. These sampling schemes, which will be discussed later in the book, have various underlying probability distributions. The most common of these being the Poisson distribution.

Consequently, a proper understanding of these probability distributions (or probability models as they are sometimes referred to) is therefore highly desirable. However, before discussing these various probability models, it would be necessary to define the concept of the moment generating functions particularly as they relate to obtaining the moments of the distributions and proving some asymptotic results.

We also introduce readers to the general theory of generalized linear models. The discussions here are brief and will help readers have an easy understanding of the theory behind categorical data analysis and some of the statistical softwares that have been developed to take advantage of this theory.

The Normal Probability Distribution

The normal or Gaussian probability distribution is referred to as the standardized normal variate and has mean zero (n=0) and variance one (a^2=1). That is, $Z \sim N(0,1)$. Categorical data analysis deals primarily with discrete data (counts or frequencies) and hence are described by discrete type distributions. We will therefore, start by examining the various distributions that we will be considering in this book. These are discussed in the next section.

The Binomial Distribution

Consider an experiment *Q* consisting of a sequence of *n* independent trials in which the outcome at each trial is binary or dichotomous. At each trial, the probability that a particular event (A)-success occurs is *n* and the probability that the event does not occur (failure) is therefore $(1-\pi)$. We shall also assume that this probability of success is constant from one trial to another. If the number of times the event A occurs in the sequence of *n* trials is represented by the random variable X, then, x is said to follow the binomial distribution with parameters *n* and π (fixed).

Example: Suppose the probability that a person suffering from migraine headache will obtain relief with a certain analgesic drug is 0.8. Five randomly selected sufferers from migraine headache are given this drug. We wish to find the following probabilities that the number obtaining relief will be:

Solution: Let *X* be the binomial random variable, with parameters $n=5$ and $\pi = 0.8$. We can usually solve the problems in (a) to (f) by the use of the cummulative distribution Function of the binomial random variable. The cdf $F(x)= P(X<x)$ for this problem is generated next with the accompanying SAS software programme.

```
DATA BINOMIAL;
DO X=0 TO 5;
P=PROBBNML(0.8,5,X);
CUM=ROUND (P.0001);
OUTPUT
END;
DROP P;
PROC PRINT NOOBS;
RUN;
```

X	CUM
0	0.0003
1	0.0067

2	0.0579
3	0.2627
4	0.6723
5	1.0000

In the programme just given, we have employed the SAS software function probbnml(p, n, X) for generating the cummulative distribution function of a binomial variable with the specified parameters. The ROUND statement asked SAS to round the probabilities to four decimal places.

Asymptotic Properties

Readers are advised to first familiarize themselves with the notation and definition adopted therein. For a binomial variate X, designated as $X\text{~}b(n, \pi)$, $0 < \pi < 1$ with π fixed, then as n –... That is, as n -> ..., the standardized binomial distribution approaches the standard normal distribution with mean 0 and variance 1. To prove this, we note from property (iii) of the moment generating functions that the results just shown are due to Cox (1970) and defined as the logit of π. On the other hand, if 0 is defined in terms of the inverse normal probability integral as 0 – \$~1(π), then 0 is in this case referred to as the *probit* of π.

6

Moments and Distribution Characteristics

DISTRIBUTION CHARACTERISTICS

The student of mechanics very early becomes familiar with the concept of "the moment of a force about a point," or the importance of the force in producing (or tending to produce) motion about an axis. In games of "seesaw," a child learning to balance another on a beam, which itself is balanced about a "fulcrum," discovers that the effectiveness of its weight on the beam is directly proportional to its distance from the fulcrum. A sketch of such a system in equilibrium is given below for reference. When equilibrium is established it is found that the product of a weight *A* and its distance a from the fulcrum *F* is equal to the similar product for *B*. These products are termed "moments" in mechanics. The two weights *A* and *B* exert their force in the same downward direction. The distances ± and ß are measured in opposite senses and therefore are appropriately given opposite signs. The two moments then are –*A*± and +*B*ß, the sum being zero when equilibrium prevails.

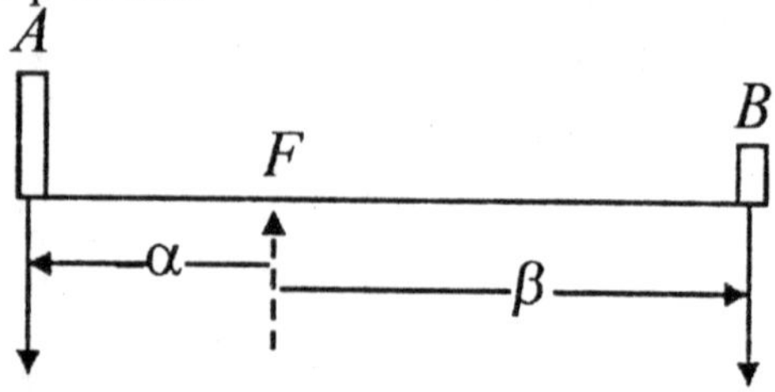

Fig. $-A\alpha$ and $+B\beta$

The effectiveness of a weight on a horizontal beam with respect to its tendency to produce motion may similarly be measured with respect to any point in the beam. The fulcrum may be moved to any point, each such point of reference being known as an *origin* of the calculated moment. If there are k objects on a weightless rod which is scaled in equal increments from some point (for example, one end, although that is immaterial), then the total moment of the system about any fixed origin a in the beam is given by the expression $f(x-a)$, where f is each weight, x is its position point on the scale, and summation is made for all k products.

Let x be the scale of a variable, of which a sample of total frequency N has been seriated into k classes, f designating the frequency of any class. A histogram drawn for this seriation may be visualized as a series of rectangles on a weightless beam, the weight of each rectangle being proportional to its area, that is, the class frequency in each case. If a is the point in the rod about which the system would be in balance if a fulcrum were placed there, then obviously a lies somewhere between the lowest class value, x_1, and the highest class value, x_k. The value of a may be determined very simply, since the condition of equilibrium demands that the total moment shall be zero. The moment of each class is $f(x - a)$, and by definition

$$f(x-a) = 0.$$

Since the only variable quantities here are f and x, the expression may be expanded to the form,

$$fx-a\,f = 0.$$

But $f = N$, and $a\,f = Na$.

Therefore $fx = Na$, or,

$$a = \frac{\sum fx}{N} = \overline{x}$$

Thus the mean of any series is the balancing point with respect to weight of the observations on the scale of measurement. The general characters of frequency distributions which have already been noted, namely, may all be defined very readily in terms of a simple series of extensions of this moment concept.

- Total frequency,

- Balancing point,
- Amount of variation,
- Skewness, and
- Kurtosis,

These extensions emanate from the basic step of considering not only the distance factor, $x-a$, but also the series of powers from zero through the fourth of this factor. Let us analyse the expression for the mean squared deviate, from which our standard measure of variation is derived. Using special frequency-table notation (x_c for class centers),

$$m_2 = \frac{\Sigma f(x_c - \overline{x})^2}{N}$$

This is simply the mean *second* moment about the origin x +. The order of the moment, designated by the qualifying adjective, is the power to which the distance factor or deviation is raised. If each value of x is considered individually the equation reduces to the general algebraic form

$$m_2 = \frac{\Sigma(x - \overline{x})^2}{N}$$

x + being reduced to the constant 1 and therefore omitted.

One may conveniently generalize the terminology for statistical moments according to the following scheme:

- The nth moment of an individual x about an origin a on its scale of measurement is $(x-a)^n$.
- The total nth moment of a series about the origin a is $f(x-a)^n$.
- The mean nth moment of a series of N individuals about an origin a is

$$\frac{\Sigma(x-a)^n}{N}.$$

The statistician in his studies of variables is concerned with average values derived from series of individuals, that is, with single figure representations of mass characteristics. The moment of an individual and the total moment of a series are but steps, as it were, to an average moment of some sort. Also he is very largely concerned for reasons of convenience with two particular origins, the mean value of the series and the scale origin of zero.

In the theoretical development of statistical concepts it is most convenient to refer measurements to the mean as origin, and as a consequence a special designation in word and symbolism has arisen for the average moments about the mean as origin. These are known as the moment coefficients, and the symbols used for them are m with definitive subscripts. Hence the previous use of m_2, meaning the second-moment coefficient, as the symbolic designation for the mean squared deviate. We shall later use m′ as the symbol for a mean moment taken about zero as origin. In accordance with this notation,

$$m'_1 = \frac{\Sigma x}{N} = \bar{x}$$

$$m'_2 = \frac{\Sigma x^2}{N}$$

and so forth.

It is interesting, though quite incidental, to note that one total moment is used as such, namely, the total zeroth moment. This is the total frequency of any series, designated herein as N. It is a derived property of numbers that any quantity when raised to the zeroth power is unity.

Thus, $(x-a)^0$ equals unity, and therefore $(x-a)^0$ equals N; the frequency of a series is a total moment, and all mean moments are therefore ratios of total moments.

One may hope that the utility of the lower moments for describing properties of frequency distributions will by now have appealed to the reader.

The zeroth defines frequency, the first defines the most broadly useful measure of type, and the second provides the standard measure of variation. When it is realised that the characteristics of skewness and kurtosis in frequency distributions may likewise be defined very simply in terms of the third and fourth-moment coefficients respectively, some appreciation of the power of analysis by means of moments will have been achieved. Before proceeding with demonstration of this, however, perhaps one might be pardoned for a digression into the historical basis for the use of the term "moment."

It was while encumbent of the professorship of applied mathematics and mechanics at University College, London, with students of engineering chiefly forming his classes, that Karl Pearson laid many of the foundations of modern statistical analysis. Challenged by visions implanted by Francis Galton masterful book *"Natural Inheritance"*, and immensely stimulated by Weldon (professor of zoology at University College) and Galton to blaze trails on the frontier of application of mathematical reasoning to the resolution of *biological* problems, Pearson faced the task of objectively defining more precisely and more completely the characteristics of frequency distributions.

In extension of the mechanical concept of moments he perceived possibilities for the development of powerful statistical tools for this purpose, possibilities which he soon developed into realities. The zeroth, first, and second moments were already in use though not recognized as such. Let us return now to a consideration of the utility of the third and fourth moments.

Kurtosis

When the even moments are considered, it will be recognized that, although the effect of weighting larger deviates through using higher powers persists, the class moments now all have the same sign. Thus the even moments cannot reflect skewness; the result is the same as if the negative half of the distribution of deviates were folded over and added to the positive half. The effect of higher even powers is to accentuate the relative importance of the extreme deviations regardless of their signs in the first power. For this reason the standard deviation is always greater than the average deviation.

If the fourth power of the deviates is used, still greater emphasis will be given in the sum to the presence of relatively long tails in a distribution, regardless of skewness. Here, then, we are dealing with relative concentration of the variates about the mean. Two distributions having the same standard deviation will have different values of m_4 if one has longer tails than the other, the higher m_4 going with the "longer-tailed" distribution.

To get such longer tails and keep the standard deviation the same, it is necessary to move frequency from the medium deviation group (which we may call the "shoulders" of the distribution), placing some individuals nearer the mean and some farther away. Thus "peakedness" at the center is increased along with tail extension when m_4 is increased without altering the *amount* of variation. It was because of this that the term *kurtosis* (implying sharpness of curvature at the center) was introduced by Karl Pearson for this feature of relative clustering about the mean.

The use of powers of deviates has introduced a method of quantitatively describing certain readily recognizable characteristics of frequency distributions. Each mean moment secured is, however, a count of units which are themselves the appropriate powers of the original unit of measurement of the variable.

Skewness and kurtosis, on the other hand, are concepts that are basically independent of any scale of measurement of a variable, and description of them should logically proceed on a scale of pure numbers. Only in this manner may degrees of both skewness and kurtosis be quantitatively described in terms which are directly comparable from one distribution to another, regardless of the scales of measurement of those variables.

Elimination of the units of measurement from moment coefficients may readily be secured by taking suitable ratios of them. Since the immediate problem is to transform the m_3 and m_4 values to pure number scales, and m_2 is the lowest moment coefficient not having a fixed value, ratios involving m_2 as the simplest reference unit would appear to offer possibilities. Karl Pearson suggested two such ratios which have proved very serviceable and are widely used. Employing Greek letters where we use the English forms, he called them the "beta coefficients." In our symbolism they may be written

$$b_1 = \frac{m_3^2}{m_2^3}, \text{ an index of skewness,}$$

and,

$$b_2 = \frac{m_4}{m_2^2}, \text{ an index of kurtosis,}$$

Both numerator and denominator of the ratio for b_1 involve sixth powers of the original unit of measurement, and therefore the coefficient is a pure number. Likewise, b_2 is a pure number derived from fourth-power units. The utility of such empirical ratios should be judged in terms of how well in practice they serve the purpose of description for which they are offered. The answer arising from experience is that basically they serve very well indeed. Certain slight modifications, aiming to introduce direction of departure from normal curve values, are preferred by some. These latter indices are designated as the "gamma coefficients" or "g statistics":

$$g_1 = \frac{m_3}{s^3}, \text{ an index of skewness,}$$

$$\text{and, } g_2 = \frac{m_4}{s^4} - 3, \text{ an index of kurtosis,}$$

Whereas the value of b_1 is zero for symmetrical distributions (since $m_3 = 0$), it is positive for all other curves because m_3 was squared to secure it. g_1, on the other hand, preserves the sign appropriate to the direction of skewness. Obviously, squaring g_1 gives b_1. Three curves having different degrees of skewness but otherwise of like character. The b and g coefficients are given in proximity to these curves, the middle curve being the symmetrical normal curve.

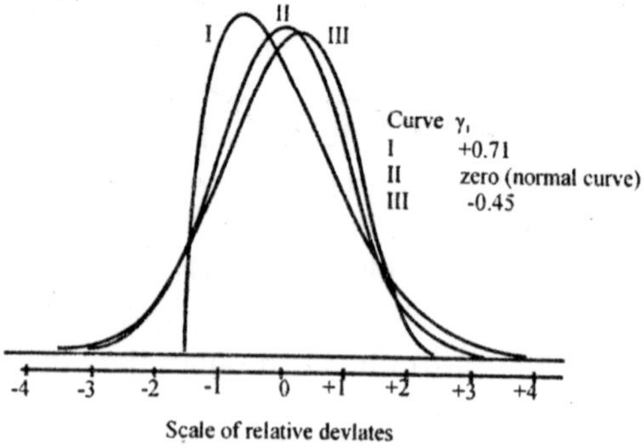

Fig. λ1 as a Measure of Skewness

The fourth-moment coefficient, m_4, is always positive, and therefore b_2 does not automatically provide a central reference value of zero. Since the normal curve is a logical central type from which departure may be measured with respect to kurtosis as well as skewness, and it has the simple whole-number value of 3 for its b_2 coefficient, kurtosis may be measured directly from this origin.

$$g_2 = b_2 - 3.$$

Thus g_2 introduces the idea of negative and positive kurtosis with respect to the normal curve as a standard of reference, positive and negative values of g_2 automatically specifying greater or less central clustering or "peakedness" than the normal curve possesses. All curves have the same mean, standard deviation, and area, the gamma coefficients changing directly with the degree of central clustering.

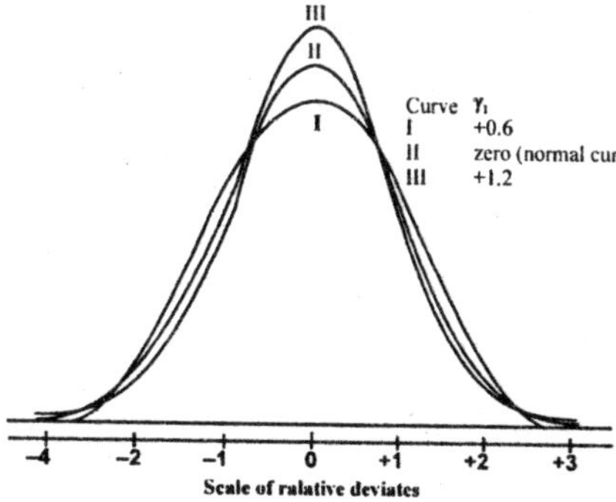

Fig. λ2 as a measure of kurtosis

Curve Fitting

The subject of fitting frequency curves to observed distributions is a very interesting one from the theoretical point of view. It has rather small practical value, however, in statistical investigations which are not of a specialized nature demanding such work. Present comment on the subject will therefore be confined to indication of the mode of procedure in what is probably the most widely used method of fitting frequency curves, that given by the Pearsonian system. When it is desired to fit a frequency curve to an observed frequency distribution a decision must be made as to the particular respects in which the curve *must fit*.

That is, a curve is fitted to a histogram by selecting an equation for the curve which has certain properties identical with corresponding properties of the histogram.

The larger the number of these identical values, the better the fit will be in general, but a curve may be "overfitted" in the sense that the curve itself reproduces the histogram too well, preserving rather than smoothing out its errors of sampling.

Thus a compromise is always necessary; as few identities are made as possible to secure reproduction of the essential features without overemphasizing small details. The system of moments leads very nicely to the establishment of descriptions of distribution characteristics on a broad basis, and Pearson has amplified their usefulness by elaborating a scheme of curves covering a wide array of combinations of these moments to the fourth order.

Starting with a differential equation which yields the normal curve as one of its solutions, Pearson derived a series of frequency functions which appeared to be adaptable to nearly all observed types of frequency distribution. The two beta coefficients already defined, together with mean, standard deviation, and total frequency, are all the values needed from an observed distribution in order to select an appropriate curve from Pearson's series for graduation purposes.

The technical discussion of procedure in fitting is quite beyond the scope of this book. The interested reader may be referred for such detail to Elderton and to Pearson's summary of procedure with most useful graphs for selecting the appropriate type of curve.

The Normal Curve

At the close of the second chapter the statement was made that biological variables for the most part do not deviate very markedly from the "normal curve" or "law of error" form.

Both Quetelet and Galton, working with anthropometric variables, were deeply impressed by the way this curve seemed to graduate their data.

The prominence they gave it probably contributed more than anything else to an early notion that this curve (which Galton often referred to as the "normal law") was the normal type for biological variables, whereas we now recognize it only as a central type in a general scheme of biological frequency distributions.

The designation of "normal curve" was accepted by Pearson as a convenient way of avoiding the issue of choosing between the originator titles "Gaussian curve," "Laplacian curve," and "Gauss Laplace curve," to which must now be added "DeMoivrian curve."

In adhering to the term *normal curve* herein, one wishes to warn the reader against any implication that other types of curves are "abnormal" in the usual sense of the word.

While holding considerable interest as a graduating function that works remarkably well for many biological variables, the normal curve is far more important for its portrayal of the way in which means and other statistics vary through errors of random sampling.

It commands so much interest that a somewhat more detailed analysis of its nature than has so far been presented seems appropriate at this point.

Mathematical equations defining the normal curve may all be reduced to the simple form,

$$w = Ce^{-\frac{1}{2}k^2} = \frac{C}{e^{\frac{1}{2}k2}}$$

where,

w = the ordinate at any point x,

C = a constant of the curve,

e = the natural logarithm base, 2.7183..., and

k = the relative deviate of x from the mean of the curve.

Even this equation may look complex enough to the biologist on first inspection, but if he is willing to refresh himself a little on exponents he may analyse the form of the curve without difficulty. First of all, let us consider the transformation from the variate scale, x, to the *relative deviate scale, k*.

The largeness or smallness of any value x may be expressed on the universal scale of pure numbers by securing its relative deviate value.

This is simply the *number of standard deviations* that x is away from the mean of the series, either in the positive or negative direction.

$$k = \frac{x - \bar{x}}{S_x}$$

Both $x–x$ + and s_x. are dimensions on the scale of measurement of x. If we divide one by the other the original units of measurements (inches, pounds, et cetera) vanish, and k as defined above becomes a pure number.

This is a generalizing transformation of scale of great importance in statistics. It enables one to compare the largeness (or smallness) of a variate in its distribution with the same property for any other variate, whatever the variable may be.

The transformation will be of great value to our development of the notion of correlation in the following chapter. At the present moment, however, it serves to place a standard scale under *all* normal distributions, permitting our study of the properties of the curve as such.

Once the mean and standard deviation of a variable have been secured, the universal scale of relative deviates may be placed in juxtaposition with its distribution.

This k scale will have its 0 value against x +, the +1 and – 1 values respectively at the deviations of plus and minus one standard deviation s_x, from x +, the scale extending without limit by such steps in both directions.

The natural base of logarithms, *e*, is of no consequence to us at the moment beyond recognition of the fact that it is a pure number greater than unity. Some knowledge is necessary, however, of the numbers which result when such a number is raised to powers.

Since the zeroth power of any number is 1, and the first power of *e* is 2.7183, then any power of e between the zeroth and the first yields a number between 1 and 2.7183. Powers of e above the first yield numbers greater than 2.7183, of course, and all positive powers of *e* will therefore yield numbers greater than 1. We are now in a position to analyse the form of the normal curve as it is defined in equation.

Ordinates of the Curve

The ordinate of the normal curve will reach its *greatest* value *C* when $E^{1/2k2}$ reaches its minimum value of unity. This minimum value occurs only when *k* is zero, corresponding to the mean value on the scale of any distribution of reference. Since *k* is squared in equation the ordinate at any chosen negative value of *k* must equal the ordinate at the same positive value of k and be less than the value *C* at the center. Therefore the distribution curve is of the symmetrical form.

Those familiar with logarithms may readily calculate a series of ordinates and draw the curve. Arbitrarily making the mid-ordinate 100, one finds the ordinates at 1, 2, and 3 standard deviations each way to be 60.7, 13.5, and 1.1 units approximately. Larger values of *k* than 3 will give very small values of *w*, which, nevertheless, will be real; so the curve has a theoretically infinite range in both directions from the mean. This does not invalidate the curve as applied to biological data, for it may readily be seen that the probability of extreme deviates in normal distributions approaches infinitesimal magnitudes quite in accord with statistical experience of biological variables.

A plot of the normal frequency function scaled in terms of relative deviates is given in panel (A). Ordinates have been drawn at distances of 1, 2, and 3 standard deviations on either side of the mean. It will be noted immediately that the curve appears to merge into the base line at, or a trifle beyond, 3 standard deviations.

At the points *k* equals +3 and –3, the height of the ordinate is only one-ninetieth part of that at the center, and the area in each tail so truncated is only a little over one-thousandth part of the whole area. For practical purposes the tangible range of the normal curve is accordingly often referred to as being roughly three standard deviations *on both sides of the mean.* As was noted above, the range is theoretically infinite, but the parts beyond ±3 standard deviations are barely perceptible.

Probabilities of the Curve

Emphasis was given to the point that, for continuous variables, frequency is properly portrayed by area. The frequency curve being the appropriate form of graduation for such variables, it follows at once that the area between any two ordinates, when expressed as a proportion of the entire area of the curve, defines the probability of occurrence of values between the two points on the scale at which those ordinates are drawn.

This is a very important concept in statistics-indeed, the one upon which the final statistical verdict in many analyses rests. The normal curve plays so crucial a part in determining such verdicts that it is important for the student to acquaint himself somewhat with the probability zones of the curve.

It is partly for this reason that the ordinates establishing such zones have been drawn in panel (A). By means of a planimeter or other suitable device the areas between the ordinates in the figure could be determined. If these were then expressed as proportions of the total area of the curve, it would be found, for instance, that a little more than two-thirds of the total area lies between the ordinates erected at ±1 standard deviation, and approximately 95 per cent lies between ±2 standard deviations.

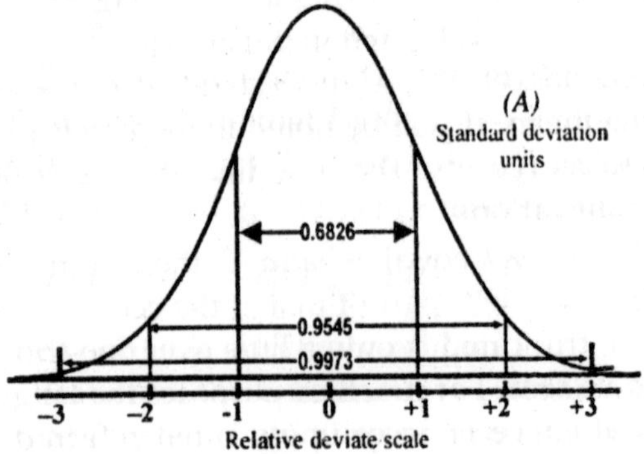

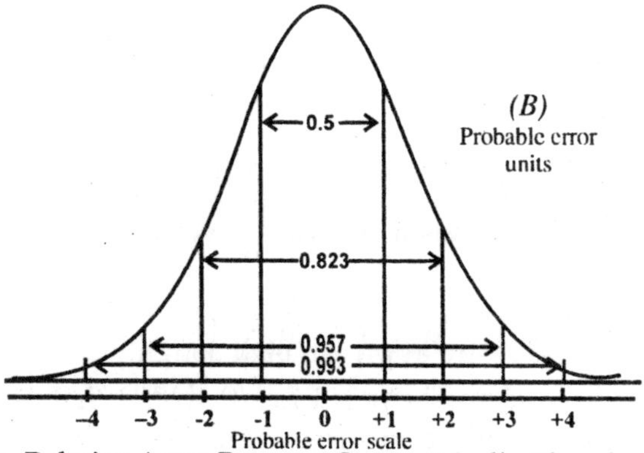

Fig. Relative Areas Between Symmetrically Placed Ordinates for the Normal Curve

These proportionate areas may be evaluated more exactly, indeed to any required degree of precision, by means of the integral calculus, and we shall rely now upon determinations so made. For the central zones indicated the values are given to four places of decimals. Note that 99.73 (or more precisely 99.73002...) per cent of the area lies within ±3 standard deviations; that is, the probability of a normally distributed variable having deviates exceeding ±3 standard deviations is slightly less than 27 in 10,000.

The ordinates embracing the central 95 per cent of cases, or truncating the extreme 2.5 per cent of the total area in each direction, have come to play an important part in statistical interpretation. These ordinates arise at the k values of ±1.9600..., or what is roughly called two standard deviations on either side of the mean. That is, in a normally distributed variable, there is only 1 chance in 20 that values will deviate from the mean by more than 1.96 standard deviations; the chance is actually 1 in 22 that values will deviate from the mean by more than two standard deviations.

The question may be raised: At what distance from the mean would the two ordinates embracing the central 50 per cent of the area be placed? Such ordinates will define the zone within and outside of which it is equally probable that individuals will occur. Those ordinates arise at the points k equals +0.6744898.... The distance of these ordinates from the mean was extensively used by astronomers in describing the amount of error made in measuring the stellar universe. Later adopted by the biologists as a measure of variation, the quantity ±0.6745 sx has become commonly known as the probable error. The reader may note that the probable-error points correspond to the first and third quartile values.

Panel (B) shows a normal curve divided into zones by ordinates raised at ±1, ±2, and ±3 probable errors. Since the factor 0.6745 differs only very slightly from two-thirds, the "5 per cent points" are approximated roughly by three probable error units, just as they are by two standard deviation units. These approximations are very useful to remember as simple guiding values in judging the importance of a deviation.

Tables of the Normal Curve

One's readily available information on probabilities in general for the normal curve needs to be much more detailed than these few skeleton values. A very frequently recurring problem in statistical analysis, especially when the variation is attributed to "errors of random sampling," is to ascertain with reasonable precision the probability that a specified value in a normal distribution will be exceeded. Occasionally, also, the normal curve ordinates are desired for plotting the curve. Accordingly, tables of these two functions are very common indeed.

Such tables need to be in a generalized form so that they may be adapted to any normal distribution with its specified values of the mean and standard deviation. The most convenient arrangement is to give the ordinates and fractional areas for the curve in terms of k, the relative deviate, as a scale of measurement, assigning a total area of one unit to the curve. Just as the use of relative deviates fixes the base scale, so the requirement that the area under the curve shall be unity fixes the vertical scale. Very full tables of ordinates and areas for the normal curve are so given in Part I of Pearson's "Tables for Statisticians and Biometricians," and those concerned with fitting the normal curve or securing probabilities with considerable precision are referred to that source.

In determining the probability that an event may be expected to occur through "errors of random sampling," one is often quite satisfied with three or four places of decimals in that probability.

The ordinates are also given in juxtaposition to three places of decimals. The area or probability, P, is that in the tail of the curve beyond the point k, expressed as a proportion of the area of half the curve. This is the same probability as that of both tails combined in relation to the whole curve.

Applications of the Normal Curve to Observed Distributions

Although it is interesting to fit the appropriate curve to any sufficiently large series of variates, it is not usually of any particular value to do so in practical analysis of data. Exercises in fitting a normal curve, therefore, become of academic interest rather than scientific importance. The situation is very different, of course, when one needs to consider the probabilities of occurrences obeying the normal law, the answers to these propositions being derived from the tables of the normal curve areas.

Those engaged in precise measurement may well follow with very keen interest the probability of the occurrence of a chance error of any given magnitude. If the errors are unbiased and normally distributed, knowledge of the mean and standard deviation of a sufficiently large series of repetitions of the measurement will give the key at once to those probabilities.

In this connection the reader may choose to apply himself to such questions with respect to the width of the spectral-line data given. The mean and standard deviation of width of line for the measurements are 178.9 and 3.583 microns, respectively.

A problem of not infrequent occurrence in educational work is that of converting scores on aptitude or intelligence tests to their equivalent values on some more or less standard score scale, or to their equivalent values on the scale of any other specified test of like purpose. We may illustrate the underlying reasoning appropriate to such transfers in terms of application of the normal curve. The distribution of scores on,

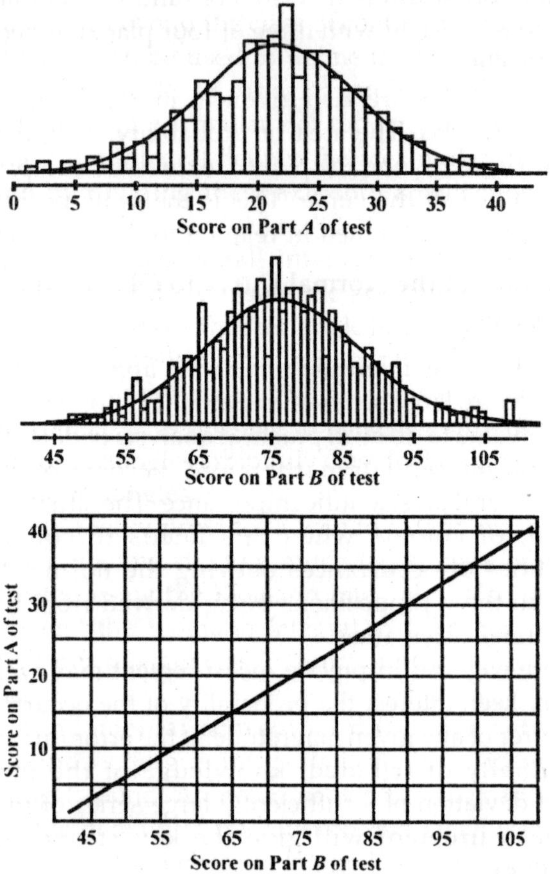

Fig. Determination of Equivalent Scores in Two Examinations by Equating the Relative Deviates

Part *B* (current affairs) of a "reflective thinking" examination given to 449 students has been used as an illustration of the commonly occurring normal distribution of measurements of specific intellectual achievements. They are reproduced in panel (B), wherein the histogram follows the original marks without grouping. The dis tribution of scores for the same students on Part *A* (science) of the examination is given in like form as panel (*A*). In both, a normal curve graduates the data satisfactorily. The actual scores on Part *A* varied from 1 to 39; those on Part *B* covered the much wider range from 47 to 108. For each score on Part *A*, what is the comparable score on Part *B* as far as performance of the group as a whole is concerned?

Equating the possible ranges of marks, 0 to 49, and 0 to 123, respectively, would not answer this question, for that would assume the two examinations to be of equal difficulty. Equating the observed ranges of scores would also fail to answer the question, for it would ignore all marks but the extreme ones—an obvious blunder. One may solve the problem appropriately only by equating those scores which are of equivalent probability of occurrence as indicated by the graduating curves. Since both curves are of the normal type, this is a simple problem; one needs only to find the scores for each test which have the same relative deviate values. The two distributions have been drawn in Fig. 16 with their means and standard deviations in correspondence. This gives to the corresponding points on the two scales equivalent probability according to the graduating curves.

Panels (*A*) and (*B*) have been prepared solely to demonstrate the underlying principle. They are by no means necessary for solution of the problem. A simple graph with one score scale along the aids of abscissas and the other along the axis of ordinates might have been prepared. A straight line drawn across the surface and passing through the points (x + 3) and ($x^+ + s_x$, $3 + s_y$)would establish the equivalent scores immediately. Such a graph forms panel (C), from which the equivalent scores may be read off with ease.

It may be noted that the problem of finding scores of equivalent probability of occurrence is simplified in the above illustration because both sets of scores are normally distributed. The same procedure might also be followed when both distributions have the same basic form as evidenced by their "gamma" coefficients. However, distributions of essentially different form would necessitate rather involved calculation to find the equivalent probability scores. Consideration of the use of the normal curve in the interpretation of the significance of differences between means and other descriptive statistics must be deferred until discussion of the "errors of random sampling" is undertaken in a later chapter. Since acquaintance with the correlation coefficient will prove most important to adequate development of the principles involved in such tests, attention will first be directed to the description of associated variation.

7

Measures of Central Tendency

There are two types of data that can be manipulated with statistical tools: cross-sectional data and time-series data. *Cross-sectional data* is a snapshot that is collected at a single point in time. It has no time dimension. A typical example of cross-sectional data would be a set of height measurements of children of different ages. Computing the average of all the measurements for a given age gives you the average height of a child of that age. The specific date and time on which any given measurement was taken is irrelevant. All that matters is the height of each child and the child's age when that height was measured.

The second type of data that you'll encounter is time-series data. With *time-series data,* every measurement has a time dimension associated with it. For example, you can look at stock-market prices over some period of time, and you'll find that the share price for any given company differs from day to day. And, not only from day to day, but also typically from hour to hour and minute to minute. A share price by itself means nothing. It's the share price coupled with the time at which that price was effective that is meaningful. A list of such prices and times over a given period is referred to as a time series.

We need to define five more terms before proceeding into the sections that describe the calculations behind the various statistics discussed in this chapter. These terms are population, sample, sample size, case, and value.

Consider the problem of having a large box of oranges for which you need to determine the sweetness level so that you can sell the box at the best possible price. That box of oranges represents your *population*. Now you could taste each orange in the box to check the sweetness level, but then you wouldn't have any oranges left to sell. Instead, you might choose to test only a few oranges—say, three. Those oranges that you test represent a *sample*, and the number of oranges in the sample represents the *sample size*. From a statistical point of view, you are testing a representative sample and then extrapolating the results to the oranges remaining in the box.

Continuing with our example, each orange in a sample represents a specific *case*. For each case—or orange—you must test the sweetness level. The result of each test will be a *value*—usually a number—indicating how sweet the orange really is. We use the term *case* to refer to a specific measurement in the sample. We use the term *value* to refer to a distinct numeric result. A value may occur more than once in a sample. For example, three oranges (three cases) may all have the same sweetness value.

Now that you have a good grasp of the statistical terms used in this chapter, you can proceed to the following sections describing various statistical measurements and calculations. Whenever you collect data, you end up with a group of scores on one or more variables.

If you take the scores on one variable and arrange them in order from lowest to highest, what you get is a distribution of scores.

Researchers often want to know about the characteristics of these distributions of scores, such as the shape of the distribution, how spread out the scores are, what the most common score is, and so on. One set of distribution characteristics that researchers are usually interested in is central tendency. This set consists of the mean, median, and mode.

STRANGE SYMBOLS USED

The Σ symbol is the summation operator. It's used as a shorthand way of telling you to add up all the values in a sample set. When you see ΣX in a formula, you know that you should sum all the values in the sample set labeled *X*. Some formulas are more complex than that. For example, you might see ΣXY in a formula.

The *X* and *Y* refer to two sample sets, and this notation is telling you to multiply corresponding elements of each set together and sum the results. So, multiply the first case in *X* by the first case in *Y*. Then, add to that the result of multiplying the second case in *X* by the second case in *Y*. Continue this pattern for all cases in the samples.

Whenever you see two elements run together, you should multiply. Thus, *XY* means to multiply the values *X* and *Y* together. $n\Sigma XY$ means to multiple the value n by the value of ΣXY.

The $\sqrt{\ }$ symbol is the square-root operator. When you see $\sqrt{n}$, you need to find the value that, when multiplied by itself, yields *n*. For example, $\sqrt{9}$ is 3, because 3×3 = 9. Most calculators allow you to easily compute square roots. Another way to express 3×3, by the way, is to write it as 32. The small, raised 2 indicates that you should multiply the value by itself.

Finally, in our notation, we use a horizontal bar over a value to indicate the mean of a sample. For example: indicates that you should compute the mean of the values for all cases in the sample labeled *X*.

SUMMATION NOTATION

The Greek letter Σ (a capital sigma) is used to designate summation.

For example, suppose an experimenter measured the performance of four subjects on a memory task. Subject 1's score will be referred to as X_1, Subject 2's as X_2, and so on. The scores are shown below:

Subject		Score
1	X_1	7
2	X_2	6
3	X_3	5
4	X_4	8

The way to use the summation sign to indicate the sum of all four X's is:

$$\sum_{i=1}^{4} X_i$$

This notation is read as follows: Sum the values of X from X_1 through X_4. The index i (shown just under the Σ sign) indicates which values of X are to be summed. The index i takes on values beginning with the value to the right of the "=" sign (1 in this case) and continues sequentially until it reaches the value above the Σ sign (4 in this case). Therefore i takes on the values 1, 2, 3, and 4 and the values of X_1, X_2, X_3, and X_4 are summed (7 + 6 + 5 + 8 = 26).

In order to make formulas more general, variables can be used with the summation notation.

For example,

$$\sum_{i=1}^{N} X_i$$

means to sum up values of X from 1 to N where N can be any number but usually indicates the sample size.

Often an abbreviated form of the summation notation is used. For example, ΣX means to sum all the values of X. When only a subset of the values of X are to be summed then the full version is required.

Thus, the sum of all elements of X except the first and the last (the N'th) would be indicated as:

$$\sum_{i=2}^{N-1} X_i$$

which would be read as the sum of X with i going from 2 to N-1. Some formulas require that each number be squared before the numbers are summed.

This is indicated by:

$$\sum_{i=2}^{N} X_i^2$$

and is equal to $7^2 + 6^2 + 5^2 + 8^2 = 174$.

The abbreviated version is simply: ΣX^2. It is very important to note that it makes a big difference whether the numbers are squared first and then summed or summed first and then squared. The symbol $(\Sigma X)^2$ indicates that the numbers should be summed first and then squared. For the present example, this equals: $(7 + 6 + 5 + 8)^2 = 26^2 = 676$. This, of course, is quite different from 174.

Sometimes a formula requires that the sum of cross products be computed. For instance, if 3 subjects were each tested twice, they might each have a score on X and on Y.

Subject	X	Y
1	2	3
2	1	6
3	4	5

The sum of cross products (2 x 3) + (1 x 6) + (4 x 5) = 32 can be represented in summation notation simply as: ΣXY.

Basic Theorems: The following data will be used to illustrate the theorems:

X	Y
3	8
2	3
4	1

$\Sigma(X + Y) = \Sigma X + \Sigma Y$
$\Sigma(X + Y) = 11 + 5 + 5 = 21$
$\Sigma X = 3 + 2 + 4 = 9$
$\Sigma Y = 8 + 3 + 1 = 12$
$\Sigma X + \Sigma Y = 9 + 12 = 21$
$\Sigma aX = a\Sigma X$ (a is a constant)
For an example, let a = 2.
$\Sigma aX = (2)(3) + (2)(2) + (2)(4) = 18$
$a\ \Sigma X = (2)(9) = 18$
$\Sigma(X\text{-}M)^2 = \Sigma X^2 - (\Sigma X)^2/N$

(N is the number of numbers, 3 in this case, and M is the mean which is also equal to 3 in this case.

$\Sigma(X\text{-}M)^2 = (3\text{-}3)^2 + (2\text{-}3)^2 + (4\text{-}3)^2 = 2$
$\Sigma X^2 = 3^2 + 2^2 + 4^2 = 29$
$(\Sigma X)^2/N = 9^2/3 = 27$
$\Sigma X^2 - (\Sigma X)^2/N = 29 - 27 = 2$

MEAN

A *mean* is a type of average, often referred to as an *arithmetic average.* Given a sample set of values, you calculate the mean by summing all the values and dividing the result by the sample size. Consider the following set of height values for seven kids in a hypothetical fifth-grade class:

100 cm
100 cm
100 cm
110 cm
115 cm
120 cm
180 cm

Given these values, you compute the mean height by summing the values and dividing by the sample size of seven.

For example:

100+100+100+110+115+120+180 = 825 cm
825 cm/7 = 117.86 cm

In this case, the mean height is 117.86 cm. As you can see, the mean height may not correspond to any student's actual height.

Arithmetic Mean

The arithmetic mean is probably the most commonly used statistic in all social science research. The mean is simply the arithmetic average of a distribution of scores, and researchers like it because it provides a single, simple number that gives a rough summary of the distribution. It is important to remember that although the mean provides a useful piece of information, it does not tell you anything about how spread out the scores are (*i.e.*, variance) or how many scores in the distribution are close to the mean. It is possible for a distribution to have very few scores at or near the mean.

The arithmetic mean is what is commonly called the average: When the word "mean" is used without a modifier, it can be assumed that it refers to the arithmetic mean. The mean is the sum of all the scores divided by the number of scores. The formula in summation notation is:

$\mu = \Sigma X/N$

where μ is the population mean and N is the number of scores.

If the scores are from a sample, then the symbol M refers to the mean and N refers to the sample size. The formula for M is the same as the formula for μ.

$M = \Sigma X/N$

The mean is a good measure of central tendency for roughly symmetric distributions but can be misleading in skewed distributions since it can be greatly influenced by scores in the tail. Therefore, other statistics such as the median may be more informative for distributions such as reaction time or family income that are frequently very skewed

The sum of squared deviations of scores from their mean is lower than their squared deviations from any other number.

For normal distributions, the mean is the most efficient and therefore the least subject to sample fluctuations of all measures of central tendency.

The formal definition of the arithmetic mean is $\mu = E[X]$ where μ is the population mean of the variable X and E[X] is the expected value of X.

MEDIAN

A *median* is yet another type of statistic that you could loosely say refers to the middle value in a sample set. To be more precise, the median is the value in a sample set that has the same number of cases below it as above it. Consider again the following height data, which is repeated from the earlier section on means:

100 cm
100 cm
100 cm
110 cm

115 cm
120 cm
180 cm

There are seven cases total. Look at the 110 cm value. There are three cases with values above 110 cm and three cases with values below 110 cm. The median height, therefore, is 110 cm.

Computing a median with an odd number of cases is relatively straightforward. Having an even number of cases, though, makes the task of computing a median more complex. With an even number of cases, you simply can't pick just one value that has an equal number of cases above and below it. The following sample amply illustrates this point:

100 cm
100 cm
100 cm
110 cm
115 cm
120 cm
125 cm
180 cm

There is no "middle" value in this sample. 110 cm won't work because there are four cases having higher values and only three having lower values. 115 cm won't work for the opposite reason—there are three cases having values above and four cases having values below. In a situation like this, there are two approaches you can take. One solution is to take the lower of the two middle values as the median value. In this case, it would be 110 cm, and it would be referred to as a *statistical median*. Another solution is to take the mean of the two middle values. In this case, it would be 112.5 cm [(115 cm+110 cm)/2]. A median value computed by taking the mean of the two middle cases is often referred to as a *financial median*.

It is actually possible to find an example where there is a median with an even number of cases. Such an example is: 1,2,3,3,4,5. There is an even number of cases in the sample; however, 3 has an equal number of cases above and below it, so that is the median. This example is exceptional, though, because the two middle cases have the same value. You can't depend on that occurring all the time.

The median is the score in the distribution that marks the 50th percentile. That is, 50% per cent of the scores in the distribution fall above the median and 50% fall below it. Researchers often use the median when they want to divide their distribution scores into two equal groups (called a median split). The median is also a useful statistic to examine when the scores in a distribution are skewed or when there are a few extreme scores at the high end or the low end of the distribution.

The median is the middle of a distribution: half the scores are above the median and half are below the median. The median is less sensitive to extreme scores than the mean and this makes it a better measure than the mean for highly skewed distributions. The median income is usually more informative than the mean income, for example.

The sum of the absolute deviations of each number from the median is lower than is the sum of absolute deviations from any other number.

The mean, median, and mode are equal in symmetric distributions. The mean is typically higher than the median in positively skewed distributions and lower than the median in negatively skewed distributions, although this may not be the case in bimodal distributions.

Computation of Median

When there is an odd number of numbers, the median is simply the middle number. For example, the median of 2, 4, and 7 is 4.

When there is an even number of numbers, the median is the mean of the two middle numbers. Thus, the median of the numbers 2, 4, 7, 12 is (4+7)/2 = 5.5.

MODE

A *mode* is another type of statistic that refers to the most frequently occurring value in a sample set. Look again at the numbers in the previous section on means. You'll see that the number of occurrences is as follows:

100 cm three
110 cm one
115 cm one
120 cm one
180 cm one

It's obvious that the most frequently occurring value is 100 cm. That value represents the mode for this sample set. The mode is the least used of the measures of central tendency because it provides the least amount of information. The mode simply indicates which score in the distribution occurs most often, or has the highest frequency.

The mode is the most frequently occurring score in a distribution and is used as a measure of central tendency. The advantage of the mode as a measure of central tendency is that its meaning is obvious. Further, it is the only measure of central tendency that can be used with nominal data. The mode is greatly subject to sample fluctuations and is therefore not recommended to be used as the only measure of central tendency. A further disadvantage of the mode is that many distributions have more than one mode. These distributions are called "multi modal." In a normal distribution, the mean, median, and mode are identical.

STANDARD ERROR

A *standard error* is a measurement of a statistical projection's accuracy based on the sample size used to obtain that projection. Suppose that you were trying to calculate the average height of students in the fifth grade. You could randomly select seven children and calculate their average height. You could then use that average height as an approximation of the average height for the entire population of fifth graders. If you select a different set of seven children, your results will probably be slightly different. If you increase your sample size, the variation between samples should decrease, and your approximations should become more accurate. In general, the greater your sample size, the more accurate your approximation will be. You can use the sample size and the standard deviation to estimate the error of approximation in a statistical projection by computing the *standard error of the mean.*

The formula to use is:

$$\frac{s}{\sqrt{n}}$$

where:

s Is the standard deviation in the sample.

n Is the number of cases in the sample.

We can use the data shown in the previous section on standard deviation in an example here showing how to compute the standard error. Recall that the sample size was seven elements and that the standard deviation was 28.56.

The standard error, then, is calculated as follows:

$$\frac{28.56}{\sqrt{7}} = \frac{28.56}{2.65} = 10.7$$

Double the sample size to 14 elements, and you have the following:

$$\frac{28.56}{\sqrt{14}} = \frac{28.56}{3.74} = 7.6$$

A standard-error value by itself doesn't mean much. It only has meaning relative to other standard-error values computed for different sample sizes for the same population. In the examples shown here, the standard error is 10.78 for a sample size of 7 and 7.64 for a sample size of 14. A sample size of 14, then, yields a more accurate approximation than a sample size of 7.

A WORD ABOUT POPULATIONS AND SAMPLES

There are two different symbols used for the mean, *X* and μ. Two different symbols are needed because it is important to distinguish between a statistic that applies to a sample and a parameter that applies to a population. A population is an individual or group that represents all the members of a certain group or category of interest. A sample is a subset drawn from the larger population. For example, let us suppose that wanted to know the average income of the current full-time, tenured faculty at Harvard. There are two ways that we can find this average.

First, get a list of every full-time, tenured faculty member at Harvard and find out the annual income of each member on this list. Because this list contains every member of the group is interested in, it can be considered a population. If we have to collect these data and calculate the mean, we would have generated a parameter, because a parameter is a value generated from, or applied to, a population.

The symbol used to represent the population mean is u. Another way to generate the mean income of the tenured faculty at Harvard would be to randomly select a sub-set of faculty names from our list and calculate the average income of this sub-set.

The sub-set is known as a sample (in this case it is a random sample), and the mean that is generated from this sample is a type of statistic. Statistics are values derived from sample data, whereas parameters are values that are either derived from, or applied to, population data.

It is important to note that all samples are representative of some population and that all sample statistics can be used as estimates of population parameters. In the case of the mean, the sample statistic is represented with the symbol.

MEASURES OF CENTRAL TENDENCY IN DEPT

The calculations for each measure of central tendency are mercifully straightforward. With the aid of a calculator or statistics software programme, you will probably never need to calculate any of these statistics. But for the sake of knowledge and in the event you find yourself without a calculator and in need of these statistics, here is the information you will need.

Because the mean is an average, calculating the mean involves adding, or summing, all of the scores in a distribution and dividing by the number of scores. So, if you have 10 scores in a distribution, you would add all of the scores together to find the sum and then divide the sum by 10, which is the number of scores in the distribution. The formula for calculating the mean is presented in Table.

Formula for calculating the mean of a distribution

$$\overline{X}, \mu = \frac{\Sigma X}{n, N}$$

where $\overline{X}$ is the sample mean

μ is the population mean
Σ means "the sum of"
X is an individual score in the distribution
n is the number of scores in the sample
N is the number of scores in the population

The calculation of the median for a simple distribution of scores is even simpler than the calculation of the mean. To find the median of a distribution, you need to first arrange all of the scores in the distribution in order, from smallest to largest. Once this is done, you simply need to find the middle score in the distribution. If there are an odd number of scores in the distribution, there will be a single score that marks the middle of the distribution.

For example, if there are 11 scores in the distribution arranged in descending order from smallest to largest, the 6th score will be the median because there will be 5 scores below it and 5 scores above it. However, if there are an even number of scores in the distribution, there is no single middle score.

In this case, the median is the average of the two scores in the middle of the distribution (as long as the scores are arranged in order, from largest to smallest). For example, if there are 10 scores in a distribution, to find the median you will need to find the average of the 5th and 6th scores. To find this average, you simply add the two scores together and divide by two.

To find the mode, there is no need to calculate anything. The mode is simply the category in the distribution that has the highest number of scores, or the highest frequency. For example, suppose you have the following distribution of IQ test scores from 10 students: 86 90 95 100 100 100 110 110 115 120

In this distribution, the score that occurs most frequently is 100, making it the mode of the distribution. If a distribution has more than one category with the most common score, the distribution has multiple modes and is called multimodal. One common example of a multimodal distribution is the bimodal distribution. Researchers often get bimodal distributions when they ask people to respond to controversial questions that tend to polarize the public. For example, if we were to ask a sample of 100 people how they feel about capital punishment, we might get the results presented in Table. On the following scale, please indicate how you feel about capital punishment.

1———2———3———4———5

Strongly opposed — Strongly in Favour

In this example, because most people either strongly opposed or strongly supported capital punishment, we ended up with a bimodal distribution of scores.

EXAMPLE: THE MEAN, MEDIAN, AND MODE OF A SKEWED DISTRIBUTIO

When scores in a distribution are normally distributed, the mean, median, and mode are all at the same point: the centre of the distribution. In the messy world of social science, however, the scores from a sample on a given variable are often not normally distributed. When the scores in a distribution tend to bunch up at one end of the distribution and there are a few scores at the other end, the distribution is said to be skewed. When working with a skewed distribution, the mean, median, and mode are usually all at different points.

It is important to note that the procedures used to calculate a mean, median, and mode are the same whether you are dealing with a skewed or a normal distribution. All that changes are where these three measures of central tendency are in relation to each other. To illustrate, this a fictional distribution of scores is created based on a sample size of 30. Suppose that we were to ask a sample of 30 randomly selected fifth graders whether they think it is important to do well in school. Suppose, further it is asked them to rate how important they think it is to do well in school using a 5-point scale, with 1= "not at all important" and 5 = "very important" Because most fifth graders tend to believe it is very important to do well in school, most of the scores in this distribution are at the high end of the scale, with a few scores at the low end. We have arranged our fictitious scores in order from smallest to largest and get the following distribution:

1	1	1	2	2	2	3	3	3	3
4	4	4	4	4	4	4	4	5	5
5	5	5	5	5	5	5	5	5	5

As you can see, there are only a few scores near the low end of the distribution (1's and 2's) and more at the high end of the distribution (4's and 5's). To get a clear picture of what this skewed distribution looks like, the graph in Figure is created.

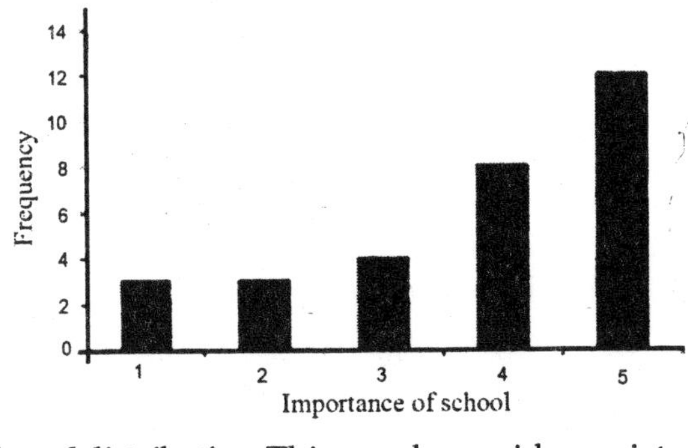

A skewed distribution: This graph provides a picture of what some skewed distributions look like. Notice how most of the scores are clustered at the higher end of the distribution and there are a few scores creating a tail towards the lower end. This is known as a negatively skewed distribution, because the tail goes towards the lower end. If the tail of the distribution were pulled out towards the higher end, this would have been a positively skewed distribution.

A quick glance at the scores in the distribution, or at the graph, reveals that the mode is 5 because there were more scores of 5 than any other number in the distribution.

To calculate the mean, we simply apply the formula mentioned earlier. That is, we add up all of the scores (EX) and then divide this sum by the number of scores in the distribution (n). This gives us a fraction of 113/30, which reduces to 3.7666. When we round to the second place after the decimal, we end up with a mean of 3.77. To find the median of this distribution, we arrange the scores in order from smallest to largest and find the middle score. In this distribution, there are 30 scores, so there will be 2 in the middle. When arranged in order, the 2 scores in the middle (the 15th and 16th scores) are both 4. When we add these two scores together and divide by 2, we end up with 4, making our median 4.

As mentioned earlier, the mean of a distribution can be affected by scores that are unusually large or small for a distribution, sometimes called outliers, whereas the median is not affected by such scores. In the case of a skewed distribution, the mean is usually pulled in the direction of the tail, because the tail is where the outliers are. In a negatively skewed distribution, such as the one presented previously, we would expect the mean to be smaller than the median, because the mean is pulled towards the tail whereas the median is not. In our example, the mean (3.77) is somewhat lower than the median (4). In positively skewed distributions, the mean is usually somewhat higher than the median.

Measures of central tendency, particularly the mean and the median, are some of the most used and useful statistics for researchers. They each provide important information about an entire distribution of scores in a single number. For example, we know that the average height of a man in the United States is five feet nine inches tall. This single number is used to summarize information about millions of men in this country. But for the same reason that the mean and median are useful, they can often be dangerous if we forget that a statistic such as the mean *ignores* a lot of information about a distribution, including the great amount of variety that exists in many distributions. Without considering the variety as well as the average, it becomes easy to make sweeping generalizations, or stereotypes, based on the mean.

8

Measures of Variability

Measures of central tendency, such as the mean and the median, provide useful information. But it is important to recognize that these measures are limited and, by themselves, do not provide a great deal of information. To illustrate, consider this example: Suppose we gave a sample of 100 fifth-grade children a survey to assess their level of depression. Suppose further that this sample had a mean of 10.0 on our depression survey and a median of 10.0 as well.

All we know from this information is that the mean and median are in the same place in our distribution, and this place is 10.0. Now consider what we do not know. We do not know if this is a high score or a low score. We do not know if all of the students in our sample have about the same level of depression or if they differ from each other. We do not know the highest depression score in our distribution or the lowest score. Simply put, we do not yet know anything about the dispersion of scores hi the distribution. In other words, we do not yet know anything about the variety of the scores in the distribution.

There are three measures of dispersion that researchers typically examine: the range, the variance, and the standard deviation. Of these, the standard deviation is perhaps the most informative and certainly the most widely used.

RANGE

The range is simply the difference between the largest

score (the maximum value) and the smallest (the minimum value) score of a distribution. This statistic gives researchers a quick sense of how spread out the scores of a distribution are, but it is not a particularly useful statistic because it can be quite misleading. For example, in our depression survey described earlier, we may have 1 student score a 1 and another score a 20, but the other 98 may all score 10. In this example, the range will be 19 (20 – 1 = 19), but the scores really are not as spread out as the range might suggest. Researchers often take a quick look at the range to see whether all or most of the points on a scale, such as a survey, were covered in the sample.

Another common measure of the range of scores in a distribution is the interquartile range (IQR). Unlike the range, which is the difference between the largest and smallest score in the distribution, the IQR is the difference between the score that marks the 75th percentile (the third quartile) and the score that marks the 25th percentile (the first quartile). If the scores in a distribution were arranged in order from largest to smallest and then divided into groups of equal size, the IQR would contain the scores in the two middle quartiles.

VARIANCE

The variance provides a statistical average of the amount of dispersion in a distribution of scores. Because of the mathematical manipulation needed to produce a variance statistic (more about this in the next section), variance, by itself, it not often used by researchers to gain a sense of a distribution. In general, variance is used more as a step in the calculation of other statistics (*e.g.*, analysis of variance) than as a stand-alone statistic. But with a simple manipulation, the variance can be transformed into the standard deviation, which is one of the statistician's favourite tools.

The variance and the closely-related standard deviation are measures of how spread out a distribution is. In other words, they are measures of variability.

The variance is computed as the average squared deviation of each number from its mean. For example, for the numbers 1, 2, and 3, the mean is 2 and the variance is:

$$\sigma^2 = \frac{(1-2)^2 + (2-2)^2 + (3-2)^2}{3} = 0.6$$

The formula (in summation notation) for the variance in a population is:

$$\sigma^2 = \frac{\sum (X-\mu)^2}{N}$$

Where μ is the mean and N is the number of scores. When the variance is computed in a sample, the statistic:

$$S^2 = \frac{\sum (X-M)^2}{N}$$

(Where M is the mean of the sample) can be used. S^2 is a biased estimate of σ^2, however. By far the most common formula for computing variance in a sample is:

$$S^2 = \frac{\sum (X-M)^2}{N-1}$$

Which gives an unbiased estimate of σ^2. Since samples are usually used to estimate parameters, s^2 is the most commonly used measure of variance. Calculating the variance is an important part of many statistical applications and analyses. It is the first step in calculating the standard deviation.

STANDARD DEVIATION

The best way to understand what a standard deviation is to consider what the two words mean. *Deviation,* in this case, refers to the difference between an individual score in a distribution and the average score for the distribution. So if the average score for a distribution is 10 (as in our previous example), and an individual child has a score of 12, the deviation is 2.

The other word in the term standard deviation is *standard.* In this case, standard means typical, or average. So a standard deviation is the typical, or average, deviation between individual scores in a distribution and the mean for the distribution.[2] This is a very useful statistic because it provides a handy measure of how spread out the scores are in the distribution. When combined, the mean and standard deviation provide a pretty good picture of what the distribution of scores is like.

In a sense, the range provides a measure of the total spread in a distribution (*i.e.*, from the lowest to the highest scores), whereas the variance and standard deviation are measures of the average amount of spread within the distribution. Researchers tend to look at the range when they want a quick snapshot of a distribution, such as when they want to know whether all of the response categories on a survey question have been used (*i.e.*, did people use all 5 points on the 5-point Likert scale?) or they want a sense of the overall balance of scores in the distribution.

Researchers rarely look at the variance alone, because it does not use the same scales as the original measure of a variable, although the variance statistic is very useful for the calculation of other statistics. The standard deviation is a very useful statistic that researchers constantly examine to provide the most easily interpretable and meaningful measure of the average dispersion of scores in a distribution.

The standard deviation formula is very simple: it is the square root of the variance. It is the most commonly used measure of spread.

An important attribute of the standard deviation as a measure of spread is that if the mean and standard deviation of a normal distribution are known, it is possible to compute the percentile rank associated with any given score. In a normal distribution, about 68% of the scores are within one standard deviation of the mean and about 95% of the scores are within two standard deviations of the mean.

The standard deviation has proven to be an extremely useful measure of spread in part because it is mathematically tractable. Many formulas in inferential statistics use the standard deviation.

Although less sensitive to extreme scores than the range, the standard deviation is more sensitive than the semi-interquartile range. Thus, the standard deviation should be supplemented by the semi-interquartile range when the possibility of extreme scores is present.

If variable Y is a linear transformation of X such that:

$$Y = bX + A,$$

then the variance of Y is:

$$\sigma_y^2 = b^2\sigma_x^2$$

where σ_x^2 is the variance of X.

The standard deviation of Y is $b\sigma_x$ where σ_x is the standard deviation of X.

MEASURES OF VARIABILITY

Standard Deviation and Variance

Standard deviations and variance are very closely related. A *standard deviation* will tell you how evenly the values in a sample are distributed around the mean. The smaller the standard deviation, the more closely the values are condensed around the mean.

The greater the standard deviation, the more values you will find at a distance from the mean.

Consider again, the following values:

100 cm
100 cm
100 cm
110 cm
115 cm
120 cm
180 cm

The standard deviation for the sample is calculated as follows:

$$\sqrt{\frac{\sum(X x)}{n\ 1}}$$

Where:

X Is the value for a specific case in the sample.
Is the mean value for all the cases in the sample.
n Is the number of cases in the sample.

Take the formula shown here for standard deviation, expand the summation for each of the cases in our sample, and you have the following calculation:

$$\sqrt{\frac{(100\ 117.86)^2+(100\quad 117.86)^2+(100\ 117.86)^2+ (115\quad 117.86)^2+(120\ 117.86)^2+(180\ 117.86)^2}{7\quad 1}}$$

Work through the math, and you'll find that the calculation returns 28.56 as the sample's standard deviation. A *variance* is nothing more than the square of the standard deviation. In our example, the standard deviation is 28.56, so the variance is calculated as follows:

$$28.56 \times 28.56 = 815.6736$$

If you look more closely at the formula for calculating standard deviation.

Calculating the Variance and Standard Deviation

There are two central issues that we need to address when considering the formulas for calculating the variance and standard deviation of a distribution:

1. Whether to use the formula for the sample or the population,
2. How to make sense of these formulas.

Sample Statistics as Estimates of Population Parameters Although researchers are sometimes interested in simply describing the characteristics (*e.g.*, mean, median, range, etc.) of a sample, for the most part we are much more concerned with what our sample tells us about the population from which the sample was drawn.

If we are conducting a study about the study habits of eighth graders, we are more concerned about eighth graders *in general* than we are about the particular group of eighth graders in our sample.

It is important to keep this in mind, because most of our statistics, although generated from sample data, are used to make estimations about the population.

This point is discussed in more detail later in the book when examining inferential statistics. But it is important to keep this in mind as you read about these measures of variation. The formulas for calculating the variance and standard deviation of sample data are actually designed to make these sample statistics better *estimates* of the population parameters (*i.e.*, the population variance and standard deviation).

It is important to note that the formulas for calculating the variance and the standard deviation differ depending on whether you are working with a distribution of scores taken from a sample or from a population. The reason these two formulas are different is quite complex, and requires more space than allowed in a short book like this. We provide an overly brief explanation here and then encourage you to find a more thorough explanation in a traditional statistics textbook. Briefly, when we do not know the population mean, we must use the sample mean as an estimate. But the sample mean will probably differ from the population mean. Whenever we use a number *other than* the actual mean to calculate the variance, we will end up with a *larger* variance, and therefore a larger standard deviation, than if we had used the actual mean. This will be true regardless of whether the number we use in our formula is smaller or larger than our actual mean.

Because the sample mean usually differs from the population mean, the variance and standard deviation that we calculate using the sample mean will probably be smaller than it would have been had we used the population mean. Therefore, when we use the sample mean to generate an *estimate* of the population variance or standard deviation, we will actually *underestimate* the size of the true variance in the population because if we had used the population mean in place of the sample mean, we would have created a larger sum of squared deviations, and a larger variance and standard deviation.

To adjust for this underestimation, we use $n - 1$ in the denominator of our sample formulas. Smaller denominators produce larger overall variance and standard deviation statistics, which will be more accurate estimates of the population parameters.

The formulas for calculating the variance and standard deviation of a population and the estimates of the population variance and standard deviation based on a sample are presented.

As we can see, the formulas for calculating the variance and the standard deviation are virtually identical. Because both require that you calculate the variance first, we begin with the formulas for calculating the variance. This formula is known as the *deviation score formula.*

When working with a population distribution, the formulas for the variance and for the standard deviation both have a denominator of *N*, which is the size of the population. In the real world of research, particularly social science research, we usually assume that we are working with a sample that represents a larger population. For example, if we study the effectiveness of our new reading programme with a class of second graders, as a researcher we assume that these particular second graders represent a larger population of second graders, or students more generally.

Because of this type of inference, researchers generally think of their research participants as a sample rather than a population, and the formula for calculating the variance of a sample is the more often used. Notice that the formula for calculating the variance of a sample is identical to that used for the population except the denominator for the sample formula is $n - 1$.

How much of a difference does it make if we use *N or n-1* in our denominator? Well, that depends on the size of the sample. If we have a sample of 500 people, there is virtually no difference between the variance formula for the population and for the estimate based on the sample. After all, dividing a numerator by 500 is almost the same as dividing it by 499. But when we have a small sample, such as a sample of 10, then there is a relatively large difference between the results produced by the population and sample formulas.

The second issue to address involves making sense of the formulas for calculating the variance. In all honesty, there will be very few times that you will need to use this formula.

Outside of our teaching duties, we haven't calculated a standard deviation by hand since our first statistics course. Thankfully, all computer statistics and spreadsheet programmes, and many calculators, compute the variance and standard deviation for us. Nevertheless, it is mildly interesting and quite informative to examine how these variance formulas work.

To begin this examination, let me remind you that the variance is simply an average of a distribution. To get an average, we need to add up all of the scores in a distribution and divide this sum by the number of scores in the distribution, which is n (remember the formula for calculating the mean. With the variance, however, we need to remember that we are not interested in the average *score* of the distribution. Rather, we are interested in the average *difference,* or *deviation,* between each score in the distribution and the mean of the distribution.

To get this information, we have to calculate a *deviation score* for each individual score in the distribution. This score is calculated by taking an individual score and subtracting the mean from that score.

If we compute a deviation score for each individual score in the distribution, then we can sum the deviation scores and divide by *n* to get the average, or standard, deviation, right? Not quite.

The problem here is that, by definition, the mean of a distribution is the mathematical middle of the distribution. Therefore, some of the scores in the distribution will fall above the mean (producing positive deviation scores), and some will fall below the mean (producing negative deviation scores). When we add these positive and negative deviation scores together, the sum will be zero.

Because the mean is the mathematical middle of the distribution, we will get zero when we add up these deviation scores no matter how big or small our sample, or how skewed or normal our distribution.

And because we cannot find an average of zero (*i.e.,* zero divided by *n* is zero, no matter what *n* is), we need to do something to get rid of this zero.

Table. Variance and Standard Deviation Formulas.

	Population	Estimate Based on a sample
Variance	$\sigma^2 = \frac{\Sigma(x-\mu)^2}{N}$	$r^2 = \frac{\Sigma(X-\overline{X})^2}{n-1}$
	Where Σ = sum X = a score in the distribution μ = the population mean N = the number of cases n in the population	Where Σ = sum X = a score in the distribution $\overline{X}$ = the sample mean = the number of cases in the sample
Standard Deviation	$\sigma = \frac{\Sigma(x-\mu)^2}{N}$	$r = \frac{\Sigma(x-\overline{x})^2}{n-1}$
	Where Σ = sum X = a score in the distribution m = the population mean N = the number of casesn in the population	Where Σ = sum X = a score in the distribution $\overline{X}$ = the sample mean = the number of cases in the sample

The solution statisticians came up with is to make each deviation score positive by squaring it. So, for each score in a distribution, we subtract the mean of the distribution and then square the deviation. If you look at the deviation score formulas in Table, we will see that all that the formula is doing with $(X - \mu,)^2$ is to take each score, subtract the mean, and square the resulting deviation score. What you get when you do this is the all-important squared deviation, which is used all the time in statistics. If we then put a summation sign in front such that we have $\Sigma(X-\mu)^2$.

What this tells us is that after we have produced a squared deviation score for each case in our distribution, we then need to add up all of these squared deviations, giving us the sum of squared deviations, or the sum of squares *(SS)*. Once this is done, we divide by the number of cases in our distribution, and we get an average, or mean, of the squared deviations. This is our variance.

The final step in this process is converting the variance into a standard deviation. Remember that in order to calculate the variance, we had to square each deviation score. We did this to avoid getting a sum of zero in our numerator. When we squared these scores, we changed our statistic from our original scale of measurement (*i.e.*, whatever units of measurement were used to generate our distribution of scores) to a squared score.

To reverse this process and give us a statistic that is back to our original unit of measurement, we merely need to take the square root of our variance. When we do this, we switch from the variance to the standard deviation. Therefore, the formula for calculating the standard deviation is exactly the same as the formula for calculating the variance, except we put a big square root symbol over the whole formula. Notice that because of the squaring and square rooting process, the standard deviation and the variance are always positive numbers.

WHY HAVE VARIANCE?

If the variance is a difficult statistic to understand, and rarely examined by researchers, why not just eliminate this statistic and jump straight to the standard deviation? There are two reasons. First, we need to calculate the variance before we can find the standard deviation anyway, so it is not more work. Second, the fundamental piece of the variance formula, which is the sum of the squared deviations, is used in a number of other statistics, most notably analysis of variance (ANOVA).

When you learn about more advanced statistics such as ANOVA, factorial ANOVA, and even regression, you will see that each of these uses the *sum of squares*, which is just another way of saying the sum of the squared deviations. Because the sum of squares is such an important piece of so many statistics, the variance statistic has maintained a place in the teaching of basic statistics.

EXAMPLE: EXAMINING THE RANGE, VARIANCE, AND STANDARD DEVIATION

We recently conducted a study in which we gave questionnaires to approximately 500 high school students in the 9th and 11th grades. In the examples that follow, we examine the mean, range, variance, and standard deviation of the distribution of responses to two of these questions.

To make sense of these (and all) statistics, you need to know exactly what the wording of the survey items was and the response scale that the students in the sample used to answer the survey items. Although this may sound obvious, we mention it here because, if you notice, much of the statistical information reported in the news (*e.g.*, the results of polls) does not provide the exact wording of the questions or the choices the sample had to respond to the questions. Without this information, it is difficult to know exactly what the responses mean, and "lying with statistics" becomes easier.

The first survey item we examine reads "If we have enough time, we can do even the most difficult work in this class." This item is designed to measure students' confidence in their abilities to succeed in their classwork. Students were asked to respond to this question by circling a number on a scale from 1 to 5. On this scale, circling the 1 means that the statement is "not at all true" and the 5 means "very true." So students were basically asked to indicate how true they felt the statement was on a scale from 1 to 5, with higher numbers indicating a stronger belief that the statement was true.

The distribution of responses produced the following statistics:

Sample Size = 491 Mean = 4.21 Standard Deviation =.98 Variance = $(.98)^2$ =.96 Range = 5-1=4

A graph of the frequency distribution for the responses on this item appears in Figure. As you can see in this graph, most of the students in the sample circled number 4 or number

5 on the response scale, indicating that they felt the item was quite true (*i.e.,* that they were confident in their ability to do their classwork if they were given enough time). Because most students circled a 4 or a 5, the average score on this item is quite high (4.21 out of a possible 5). This is a negatively skewed distribution.

The graph in Figure. also provides information about the variety of scores in this distribution. Although our range statistic is 4, indicating that students in the sample circled both the highest and the lowest number on the response scale, we can see that the range does not really provide much useful information. For example, the range does not tell us that most of the students in our sample scored at the high end of the scale. By combining the information from the range statistic with the mean statistic, we can reach the following conclusion: "Although the distribution of scores on this item covers the full range, it appears that most scores are at the higher end of the response scale."

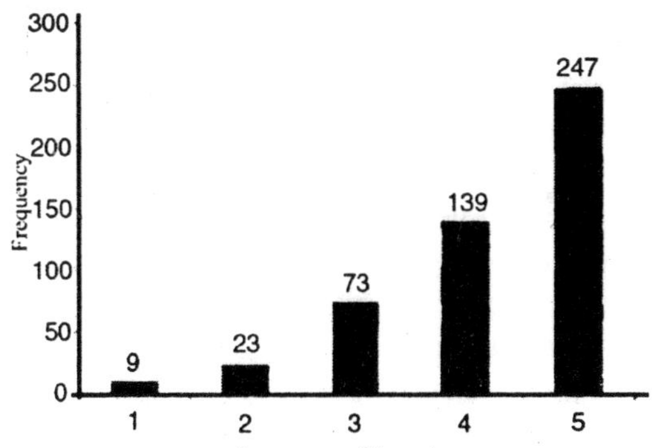

Fig. Frequency Distribution of Scores on Confidence Item.

Now that we've determined that:

- The distribution of scores covers the full range of possible scores (*i.e.,* from 1 to 5),
- Most of the responses were at the high end of the scale (because the mean is 4.21 out of a possible 5),

we may want a more precise measure of the average amount of variety among the scores in the distribution. For this we turn to the variance and standard deviation statistics. In this example, the variance (.96) is almost exactly the same as the standard deviation.

This is something of a fluke. Do not be fooled. It is quite rare for the variance and standard deviation to be so similar. In fact, this only happens if the standard deviation is about 1.0, because 1.0 squared is 1.0. So in this rare case, the variance and standard deviation provide almost the same information. Namely, they indicate that the average difference between an individual score in the distribution and the mean for the distribution is about 1 point on the 5-point scale. Taken together, these statistics tell us the same things that the graph tells us, but more precisely.

Namely, we now know that:

- Students in the study answered this item covering the whole range of response choices (*i.e.*, 1 – 5);
- Most of the students answered at or near the top of the range, because the mean is quite high;
- The scores in this distribution generally packed fairly closely together with most students circling a number within 1 point of the mean, because the standard deviation was.98.

The variance tells us that the average *squared deviation* was.96, and we scratch our heads, wonder what good it does us to know the average squared deviation, and move on.

In our second example, we examine students' responses to the item "We would feel really good if we were the only one who could answer the teacher's question in class." This item is one of several on the survey designed to measure students' desires to demonstrate to others that they are smart, or academically able.

We received responses from 491 students on this item and the distribution produced the following statistics: Sample Size = 491 Mean = 2.92 Standard Deviation = 1.43 Variance = $(1.43)^2$ = 2.04 Range = 5-1= 4

In Figure, a graph is presented that illustrates the distribution of students' responses to this item across each of the five response categories. It is obvious, when looking at this graph, how the distribution of scores on this item differs from the distribution of scores on the confidence item presented earlier. But if we didn't have this graph, how could we use the statistics to discover these differences between the distributions of scores on these two items?

Notice that, as with the previous item, the range is 4, indicating that some students circled the number 1 on the response scale and some circled the number 5. Because the ranges for both the confidence and the wanting to appear able items are equal (*i.e.*, 4), they do nothing to indicate the differences in the distributions of the responses to these two items. That is why the range is not a particularly useful statistic-it simply does not provide very much information.

Our first real indication that the distributions differ substantially comes from a comparison of the means. In the previous example, the mean of 4.21 indicated that most of the students must have circled either a 4 or a 5 on the response scale. For this second item, the mean of 2.92 is a bit less informative. Although it provides an average score, it is impossible from just examining the mean to determine whether most students circled a 2 or 3 on the scale, or whether roughly equal numbers of students circled each of the five numbers on the response scale, or whether almost half of the students circled 1 whereas the other half circled 5. All three scenarios would produce a mean of about 2.92, because that is roughly the middle of the response scale.

To get a better picture of this distribution, we need to consider the standard deviation in conjunction with the mean. Before discussing the actual standard deviation for this distribution of scores, let us briefly consider what we would expect the standard deviation to be for each of the three scenarios just described. First, if almost all of the students circled a 2 or a 3 on the response scale, we would expect a fairly small standard deviation, as we saw in the previous example using the confidence item.

The more similar the responses are to an item, the smaller the standard deviation. However, if half of the students circled 1 whereas the other half circled 5, we would expect a large standard deviation (about 2.0) because each score would be about two units away from the mean (if the mean is about 3.0 and each response is either 1 or 5, each response is about two units away from the mean). Finally, if the responses are fairly evenly spread out across the five response categories, we would expect a moderately sized standard deviation (about 1.50).

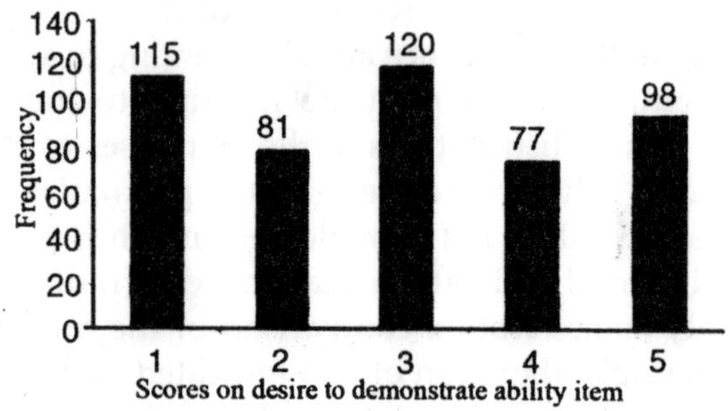

Fig. Frequency Distribution for Scores on Desire to Demonstrate Ability Item.

Now, when we look at the actual mean for this distribution (2.92) and the actual standard deviation (1.43), we can develop a rough picture of the distribution in our minds. Because we know that on a scale from 1 to 5, a mean *of 2.92* is about in the middle, we can guess that the distribution looks somewhat symmetrical (*i.e.*, that there will be roughly the same number of responses in the 4 and 5 categories as there are in the 1 and 2 categories.

Furthermore, because we've got a moderately sized standard deviation of 1.43, we know that the scores are pretty well spread out, with a healthy number of students in each of the five response categories. So we know that we didn't get an overwhelming number of students circling 3 and we didn't get students circling only 1 or 5. At this point, that is about all we can say about this distribution: The mean is near the middle of the scale and the responses are pretty well spread out across the five response categories. To say any more, we would need to look at the number of responses in each category, such as that presented in Figure.

As we look at the actual distribution of scores presented in the graph in Figure, we can see that our predictions that we generated from our statistics about the shape of the distribution were pretty accurate. Notice that we did not need to consider the variance at all, because the variance in this example (2.04) is on a different scale of measurement than our original 5-point response scale, and therefore is very difficult to interpret.

Variance is an important statistic for many techniques (*e.g.*, ANOVA, regression), but it does little to help us understand the shape of a distribution of scores.

The mean, standard deviation, and to a lesser extent the range, when considered together, can provide a rough picture of a distribution of scores. Often, a rough picture is all a researcher needs or wants. Sometimes, however, researchers need to know more precisely the characteristics of a distribution of scores. In that case, a picture, such as a graph, may be worth a thousand words.

Another useful way to examine a distribution of scores is to create a boxplot. In Figure, a boxplot is presented for the same variable that is represented in Figure, wanting to demonstrate ability. This boxplot was produced in the SPSS statistical software programme. The box in this graph contains some very useful information. First, the thick line in the middle of the box represents the median of this distribution of scores. The top line of the box represents the 75th percentile of the distribution and the bottom line represents the 25th percentile.

Therefore, the top and bottom lines of the box reveal the interquartile range (IQR) for this distribution. In other words, 50 per cent of the scores on this variable in this distribution are contained within the upper and lower lines of this box (*i.e.*, 50% of the scores are between just above a score of 2 and just below a score of 4). The vertical lines coming out of the top and bottom of the box and culminating in horizontal lines reveal the largest and smallest scores in the distribution, or the range. These scores are 5 and 1, producing a range of 5 – 1 = 4. As you can see, the boxplot in figure contains a lot of useful information about the spread of scores on this variable in a single picture.

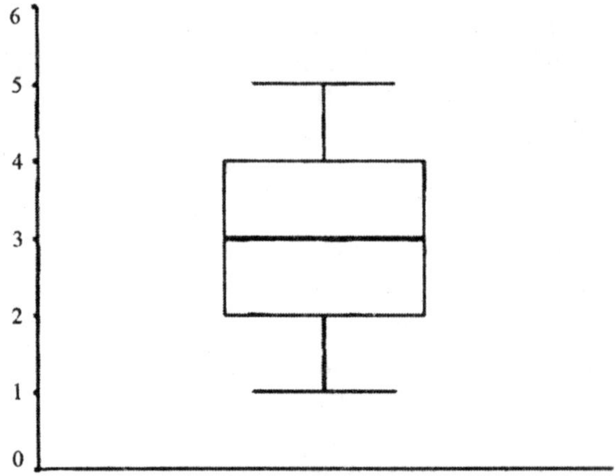

Fig. Boxplot for the Desire to Appear able Variable.

Measures of variation, such as the variance, standard deviation, and range, are important descriptive statistics. They provide useful information about how spread out the scores of a distribution are, and the shape of the distribution. Perhaps even more important than their utility as descriptors of a single distribution of scores is their role in more advanced statistics such as those coming in later chapters. We examine the properties of a theoretical distribution, the normal distribution, that has a specific shape and characteristics. We can see how the normal distribution can be used make inferences about the population based on sample data.

9

Analysis of Variance

The purpose of a one-way analysis of variance (one-way ANOVA) is to compare the means of two or more groups (the independent variable) on one dependent variable to see if the group means are significantly different from each other. In fact, if you want to compare the means of two independent groups on a single variable, you can use either an independent samples *t* test or a oneway ANOVA.

The results will be identical, except instead of producing a *t* value, the ANOVA will produce an Fratio, which is simply the *t* value squared. Because the *t* test and the one-way ANOVA produce identical results when there are only two groups being compared, most researchers use the one-way ANOVA only when they are comparing three or more groups. To conduct a one-way ANOVA, you need to have a categorical (or nominal) variable that has at least two independent groups as the independent variable and a continuous variable as the dependent variable.

Because the independent *t* test and the one-way ANOVA are so similar, people often wonder, Why don't we just use *t* tests instead of one-way ANOVAs? Perhaps the best way to answer this question is by using an example. Suppose that we want to go into the potato chip business. We've got three different recipes, but because I'm new to the business and don't have a lot of money, we can produce only one flavour. We want to see which flavour people like best and produce that one. randomly select 90 adults and randomly divide them into three groups. One group tries our BBQ-flavoured chips, the second group tries our ranch-flavoured chips, and the third group tastes our cheese-flavoured chips. All participants in each group fill out a rating form after tasting the chips to indicate how much they liked the taste of the chips. The rating scale goes from a score of 1 ("Hated it") to 7 ("Loved it"). We then compare the average ratings of the three groups to see which group liked the taste of their chips the most. In this example, the chip flavour (BBQ, Ranch, Cheese) is our categorical, independent variable and the rating of the taste of the chips is our continuous, dependent variable.

To see which flavour received the highest average rating, we could run three separate independent *t* tests comparing:

- BBQ with Ranch,
- BBQ with Cheese,
- Ranch with Cheese.

The problem with running three separate *t* tests is that each time we run a *t* test, we must make a decision about whether the difference between the two means is meaningful, or statistically significant. This decision is based on probability, and every time we make such a decision, there is a slight chance we might be wrong.

The more times we make decisions about the significance of *t* tests, the greater the chances are that we will be wrong. In other words, the more *t* tests we run, the greater the chances become of deciding that a *t* test is significant (*i.e.*, that the means being compared are really different) when it really is not. A one-way ANOVA fixes this problem by adjusting for the number of groups being compared. To see how it does this, let's take a look at one-way ANOVA in more detail.

ONE-WAY ANOVA IN DEPTH

The purpose of a one-way ANOVA is to divide up the variance in some dependent variable into two components: the variance attributable to between-group differences, and the variance attributable to within-group differences, also known as *error*. When we select a sample from a population and calculate the mean for that sample on some variable, that sample mean is our best predictor of the population mean. In other words, if we do not know the mean of the population, our best guess about what the population mean is would have to come from the mean of a sample drawn randomly from that population.

Any scores in the sample that differ from the sample mean are believed to include what statisticians call error. For example, suppose we have a sample of 20 randomly selected fifth graders. We give them a test of basic skills in math and find out that, in our sample, the average number of items answered correctly on our test is 12. If we were to select one student in our sample and find that she had a score of 10 on the test, the difference between her score and the sample mean would be considered error.

The variation that we find among the scores in a sample is not just considered error. In fact, it is thought to represent a specific kind of error: *random error*. When we select a sample at random from a population, we expect that the members of that sample will not all have identical scores on our variable of interest (*e.g.*, test scores). That is, we expect that there will be some variability in the scores of the members of the sample. That's just what happens when you select members of a sample randomly from a population. Therefore, the variation in scores that we see among the members of our sample is just considered random error.

The question that we can address using ANOVA is this: Is the average amount of difference, or variation, between the scores of members *of different* samples large or small compared to the average amount of variation *within* each sample, otherwise known as random error (a.k.a. error)? To answer this question, we have to determine three things. First, we have to calculate the average amount of variation within each of our samples.

This is called the mean square within (MS_{W}) or the mean square error (*MS e*). Second, we have to find the average amount of variation *between* the groups. This is called the mean square between (*MS b*). Once we've found these two statistics, we must find their ratio by dividing the mean square between by the mean square error. This ratio provides our *F* value, and when we have our *F* value we can look at our family of *F* distributions to see if the differences between the groups are statistically significant.

Formula for the F value:

$$F = \frac{\text{mean square between}}{\text{mean square error}}$$

or

$$F = \frac{MS_b}{MSe}$$

where F is the F value

MS_b is mean square between group

MS_e is mean square error, or within groups

Recall that the formula for calculating an independent *t* test also involves finding a ratio. The top portion of the fraction is the difference between two sample means, which is analogous to the mean square between *(MSb)* just presented.

The only differences between the two are:

- Rather than finding a simple difference between two means as in a *t* test, in ANOVA we are finding the *average* difference between means, because we often are comparing more than two means;
- We are using the squared value of the difference between the means. The bottom portion of the fraction for the *t* test is the standard *error* of the difference between two sample means.

This is exactly the same as the *average,* or standard, error within groups. In the formula used to calculate the *F* value in ANOVA, we must square this average within-group error. So, just as in the *t* test, in ANOVA we are trying to find the average difference *between* group means relative to the average amount of variation *within* each group.

To find the MS_e and MS_b, we must begin by finding the sum of squares error (SS_e) and the sum of squares between (SS_b). This sum of squares idea is not new. It is the same old sum of squares introduced back when talking about variance and standard deviation. Sum of squares is actually short for *sum of squared deviations.* In the case of ANOVA, we have two types of deviations. The first is deviations between each score in a sample and the mean for that sample (*i.e.*, error). The second type of deviation is between each sample mean and the mean for all of the groups combined, called the grand mean (*i.e.*, between groups).

To find the sum of squares error:

1. Subtract the group mean from each individual score in each group: $(X - \bar{X})$.
2. Square each of these deviation scores: $(X - \bar{X})^2$.
3. Add them all up for each group: $\Sigma(X - \bar{X})^2$.
4. Then add up all of the sums of squares for all of the groups:
   $\Sigma(X_1 - \bar{X}_1)^2 + \Sigma(X_2 - \bar{X}_2)^2 + \ldots + \Sigma(X_k - \bar{X}_k)^2$

Note: The subscripts indicate the individual groups, through the last group, which is indicated with the subscript k.The method used to calculate the sum of squares between groups *(SSb)* is just slightly more complicated than the SS_e formula.

To find the SS_b we:

1. Subtract the grand mean from the group mean: $(\bar{X} - \bar{X}_T)$; $_T$ indicates total, or the mean for the total group.
2. Square each of these deviation scores: $(\bar{X} - \bar{X}_T)^2$.
3. Multiply each squared deviation by the number of cases in the group: $[n(\bar{X} - \bar{X}_T)^2]$.
4. Add these squared deviations from each group together:

$$\Sigma[n(\bar{X} - \bar{X}_T)^2].$$

The only real differences between the formula for calculating the SS_e and the SS_b are:

1. In the SS_e we subtract the group mean from the individual scores in each group whereas in the SS_b we subtract the grand mean from each group mean.
2. In the SS_b we multiply each squared deviation by the number of cases in each group. We must do this to get an approximate deviation between the group mean and the grand mean for *each case in every group*.

If we were to add the SS_e to the SS_b, the resulting sum would be called the sum of squares total *(SS T)*. A brief word about the SS_T is in order. Suppose that we have three randomly selected samples of children. One is a sample of 5th graders, another is a sample of 8th graders, and the third is a sample of 11th graders.

If we were to give each student in each sample a spelling test, we could add up the scores for all of the children in the three samples combined and divide by the total number of scores to produce one average score. Because we have combined the scores from all three samples, this overall average score would be called the grand mean, or total mean, which would have the symbol X_T.

Using this grand mean, we could calculate a squared deviation score for each child in all three of our samples combined using the familiar formula $(X–X\ T)^2$. The interesting thing about these squared deviations is that, for each child, the difference between each child's score and the grand mean is the sum of that child's deviation from the mean of his or her own group plus the deviation of that group mean from the grand mean.

So, suppose Jimmy is in the fifth-grade sample. Jimmy gets a score of 25 on the spelling test. The average score for the fifth-grade sample is 30, and the average score for all of the samples combined (*i.e.*, the grand mean) is 35. The difference between Jimmy's score (25) and the grand mean (35) is just the difference between Jimmy's score and the mean for his group (25–30=–5) plus the difference between his group's mean and the grand mean (30–35=–5). Jimmy's deviation from the grand mean is –10. If we square that deviation score, we end up with a squared deviation of 100 for Jimmy.

Now, if we calculated a deviation score for each child in all three samples and added up all of these deviation scores using the old E $(X–X\ T)^2$ formula, the result would be the sum of squares total, or the SS_T. The interesting thing about this SS_T is that it is really just the sum of the SS_b and the SS_e. $SS\ T = SSb + SS_e$. This makes sense, because, as we saw with Jimmy, the difference between any individual score and the grand mean is just the sum of the difference between the individual score and the mean of the group that the individual is from plus the difference between that group mean and the grand mean. This is the crux of ANOVA.

DECIDING IF THE GROUP MEANS ARE SIGNIFICANTLY DIFFERENT

Once we have calculated the SS_b and the SS_e, we have to convert them to average squared deviation scores, or *MSb* and MS_e. This is necessary because there are far more deviation scores in the SS_e than there are in the SS_b, so the sums of squares can be a bit misleading. What we want to know in an ANOVA is whether the *average* difference between the group means is large or small relative to the *average* difference between the individual scores and their respective group means, or the average amount of error within each group. To convert these sums of squares into mean squares, we must divide the sums of squares by their appropriate degrees of freedom.

For the SS_b, remember that we are only making comparisons between each of the groups. The degrees of freedom for the SS_b is always the number of groups minus 1. If we use *K* to represent the number of groups, and *df* to represent degrees of freedom, then the formula for the between groups degrees of freedom is $df = K-1$. So, to convert an *SSb* to an *MSb, we* divide SS_b *by* $K- 1$.

The degrees of freedom for SS_e is found by taking the number of scores in each group and subtracting 1 from each group. So, if we have three groups, our *df for* SS_e will be $(n_1 - 1) + (n_2 - 1) + (n_3 - 1)$. Notice that this is the same formula for the degrees of freedom that was used for the independent samples *t* test.

The only difference is that we have one more group here. A simpler way to write this *df* formula is *N-K,* where N is the total number of cases for *all* groups combined and *K* is the number of groups. Once we have this *df,* we can convert the SS_e into an MS_e by simply dividing SS_e *by* $N-K$. Table contains a summary of the formulas for converting the sums of squares into mean squares.

Table. Converting Sums of Squares into mean squares

$MS_e = \frac{SS_e}{K-1}$	$MS_e = \frac{SS_e}{N-K}$
MS_e =Mean squres between groups	MS_e =Mean squres error
SS_e =Sum of squres between groups	SS_e =Sum of squres error
N=The number of cases combined across all groups	K=The number of groups
	N=The number of cases combined across all groups

Once we have found our *MSb and* our MS_e, all we have to do is divide *MSb by* MS_e to find our *F* value. Once we've found our *F* value, we need to look in our table of *F* values to see whether it is statistically significant. This table of *F* values is similar to the table of *t* values we used, with one important difference. Unlike *t* values, the significance of *F* values depends on both the number of cases in the samples (*i.e.*, the *df for* MS_e) *and* the number of groups being compared (*i.e.*, the *df for* MS_b). This second *df is* critical, because it is what is used to control for the fact that we are comparing more than two groups. Without it, we might as well conduct multiple *t* tests, and this is problematic for the reasons discussed at the beginning of the chapter. Critical values for *F* associated with different alpha levels. If our observed value of *F* is larger than our critical value of *F,* we must conclude that there are statistically significant differences between the group means.

FACTORIAL ANALYSIS OF VARIANCE

We examined one-way ANOVA. In this chapter and the two that follow, we explore the wonders of two more advanced methods of analyzing variance: factorial ANOVA and repeated-measures ANOVA. These techniques are based on the same general principles as one-way ANOVA. Namely, they all involve the partitioning of the variance of a dependent variable into its component parts (*e.g.*, the part attributable to between-group differences, the part attributable to within-group variance, or error). In addition, these techniques allow us to examine more complex, and often more interesting questions than is allowed by simple one-way ANOVA. These more advanced statistical techniques involve much more complex formulas than those we have seen previously. Therefore, in this chapter and those that follow only a basic introduction to the techniques is offered. You should keep in mind that there is much more to these statistics than described in these pages, and should consider reading more about them in the suggested readings at the end of each chapter.

WHEN TO USE FACTORIAL ANOVA

Factorial ANOVA is the technique to use when you have one continuous (*i.e.*, interval or ratio scaled) dependent variable and two or more categorical (*i.e.*, nominally scaled) independent variables. For example, suppose we want to know whether boys and girls differ in the amount of television they watch per week, on average. Suppose we also want to know whether children in different regions of the United States (*i.e.*, East, West, North, and South) differ in their average amount of television watched per week. In this example, average amount of television watched per week is our dependent variable, and gender and region of the country are our two independent variables. This is known as a 2 × 4 factorial analysis, because one of our independent variables has two levels (gender) and one has four levels (region). If we were writing about this analysis in an academic paper, we would write "we conducted a 2 (gender) × 4 (region) factorial ANOVA."

Now when we run our factorial ANOVA, we will get three interesting results. First, we will get two main effects, one for our comparison of boys and girls and one for our comparison of children from different regions of the country. These results are similar to the results we would get if we simply were to run two one-way ANOVAs, with one important difference, which we describe in the next section. In addition to these main effects, our factorial ANOVA also produces an interaction effect, or simply an interaction. An interaction is present when the differences between the groups of one independent on the dependent variable vary according to the level of a second independent variable. Interaction effects are also known as moderator effects. We discuss interactions in greater detail in the next section as well. For now, suffice it to say that interaction effects are often very interesting and important pieces of information for social scientists.

SOME CAUTIONS

Just as with one-way ANOVA, when conducting a factorial ANOVA it is important to determine whether there are roughly an equal number of cases in each group and whether the amount of variance within each group is roughly equal (known as homogeneity of variance). The ideal situation in ANOVA is to have roughly equal sample sizes in each group and a roughly equal amount of variation (*e.g.*, the standard deviation) in each group. If the sample sizes are equal but the standard deviations are significantly different, this is not a big problem. But if the sample sizes *and* the standard deviations are quite different in the various groups, there is a problem. This is particularly likely to be a problem in factorial ANOVA because the combination of independent variables often produces a small number of cases in each group.

In our previous example, suppose we have 20 boys and 20 girls in our entire sample. In addition, suppose that we have 20 children from each of the four regions in our sample. To test the main effects, these numbers are acceptable. That is, it is reasonable to compare 20 boys to 20 girls if we want to know whether boys and girls differ in their average amount of television viewing.

Similarly, it is reasonable to compare 20 children from each of the four different regions of the country. But suppose that in the West, our sample of 20 children includes only 5 girls and 15 boys, whereas in the North our sample includes 15 girls and only 5 boys. When we divide up our sample by *two* independent variables, it is easy to wind up with cell sizes that are too small to conduct meaningful ANOVAs. A "cell" is a subset of cases representing one unique point of intersection between the independent variables. In the aforementioned example, there would be eight cells: girls from the West, boys from the West, girls from the South, boys from the South, and so on. When you consider that factorial ANOVAs can have more than two independent variables, the sample can be subdivided a number of times. Without a large initial sample, it is easy to wind up with cells that contain too few cases. As a general rule, cells that have fewer than 10 cases are too small to include in ANOVAs, and cell sizes of at least 20 are preferred.

FACTORIAL ANOVA IN DEPTH

When dividing up the variance of a dependent variable, such as hours of television watched per week, into its component parts, there are a number of components that we can examine. In this section, we examine three of these components: The main effects, interaction effects, and simple effects. In addition, we also present an introduction to the idea of partial and controlled effects.

MAIN EFFECTS AND CONTROLLED OR PARTIAL EFFECTS

As mentioned earlier, a factorial ANOVA will produce main effects for each independent variable in the analysis. These main effects will each have their own *F* value, and are very similar to the results that would be produced if you just conducted a one-way ANOVA for each independent variable on the dependent variable. However, there is one glorious benefit of looking at the main effects in a factorial ANOVA rather than separate one-way ANOVAs: When looking at the main effects from a factorial ANOVA, it is possible to test whether there are significant differences between the groups of one independent variable on the dependent variable while controlling for, or partialing out the effects of the other independent variable(s) on the dependent variable. Let me clarify this confusing sentence by returning to our example of television viewing.

Suppose that when we examine whether boys and girls differ in the average amount of television they watch per week, we find that there is a significant difference: Boys watch significantly more television than girls. In addition, suppose that children in the North watch, on average, more television than children in the South.

Now, suppose that, in our sample of children from the Northern region of the country, there are twice as many boys as girls, whereas in our sample from the South there are twice as many girls as boys. Now I've got a potential problem. How do we know whether our finding that children in the North watch more television than children in the South is not just some *artifact* caused by the greater proportion of boys in our Northern sample? By "artifact" we mean that the North-South difference is merely a *by-product* of the difference between boys and girls; region of the country is not an important factor in and of itself.

Think about it: If we already know that boys watch more television, on average, than girls, then we would *expect* our Northern sample to watch more television than our Southern sample because there is a greater proportion of boys in our Northern sample than in the Southern sample. So our question is this: How can we determine whether there is a difference in the average amount of television watched by children in the North and South *beyond* the difference caused by the unequal proportions of boys and girls in the samples from these two regions. Phrased another way, is there an effect of region on television viewing *beyond,* or *in addition to* the effect of gender?

To answer this intriguing question, we must examine the main effect of region on television viewing after *controlling for,* or *partialing out* the effect of gender. We can do this in a factorial ANOVA. To understand how this is accomplished, keep in mind that what we are trying to do with an ANOVA is to *explain* the variance in our dependent variable (amount of television children watch per week) by dividing that variance up into its component parts.

If boys and girls differ in how much they watch television, then part of the variance is explained, or accounted for, by gender. In other words, we can understand a bit of the differences among children in their weekly television viewing if we know their gender. Now, once we remove that portion of the total variance that is explained by gender, we can test whether any *additional* part of the variance can be explained by knowing what region of the country children are from. If children from the North and South *still* differ in the amount of television they watch, after *partialing out* or *controlling for* the chunk of variance explained by gender, then we know that there is a main effect of region *independent* o/the effect of gender.

In statistical jargon, we would say "There is a main effect of region on amount of television watched *after controlling for the effect of gender.*" This is powerful information. In factorial ANOVA, it is possible to examine each main effect and each interaction effect when controlling for *all* other effects in the analysis.

INTERACTIONS

A second benefit of factorial ANOVA is that it allows researchers to test whether there are any statistical interactions present. Interactions can be a complex concept to grasp. Making the whole issue even more confusing is that the level of possible interactions increases as the number of variables increases.

For example, when there are two independent variables in the analysis, there are two possible main effects and one possible two-way interaction effect (*i.e.,* the interaction between the two independent variables). If there are three independent variables in the analysis, there are three possible main effects, three possible two-way interaction effects, and one possible three-way interaction effect. The whole analysis can get very complicated very quickly. To keep things simple, let's take a look at two-way interactions first.

In our television-viewing example, suppose that we randomly selected 25 boys and 25 girls from each of the four regions of the country, measured the number of hours each child spent watching television, and calculated the averages for each group.

Note: Unlike the example provided earlier, in this example there are equal numbers of boys and girls from each region in the sample. These averages are presented in Table.

Table: Mean Amounts of Television Viewed by Gender and Region

	North	East	West	South	Overall average by Gender
Girls	20	15	15	10	15
Boys	25	20	20	25	22.5
Overall average by region	22.5	17.5	17.5	17.5	

As we see when examining the means in Table, boys in each region watch more television, on average, than girls. The overall averages by gender presented in the last column indicate that there appears to be a main effect for gender, with boys watching an average of 22.5 hours of television per week and girls watching an average of only 15 hours per week. When we look at the overall averages presented for each region (bottom row), we can see that children in the North watch more television, on average, than do children in the other three regions.

Therefore, we can tell that there appear to be main effects for gender and region on amount of television watched. Notice that it is said *"appear* to be main effects." To determine whether these main effects of statistically significant, we would have to determine the probability of obtaining differences of this size between randomly selected groups of this size.

Once we have examined the main effects, we can turn our attention to the possible interaction effects. To do this, we need to examine the means in each of the eight cells presented in Table (*i.e.*, Northern boys, Northern girls, Eastern boys, Eastern girls, etc.). When we examine these means, we can see that in the North, East, and West, boys watch an average of 5 hours more television per week than girls. But in the South, boys watch an average of 15 more hours of television than girls. Therefore, it appears that the differences in the amount of television watched by girls and boys are not uniform across the four regions of the country. In other words, the relationship between gender and amount of television watched *depends on,* or *is moderated by,* the region of the country. Because the definition of a two-way interaction is that the relationship between an independent variable and a dependent variable is moderated by a second independent variable, we appear to have a two-way interaction here.

When we find a statistically significant interaction (again, we must examine the *p* value of the *F* ratio for the interaction term to determine if it is statistically significant), we must determine the *nature* of the interaction and then *describe* the interaction. One excellent method for getting a handle on the nature of the interaction is to depict it graphically. To do this, all we need to do is to graph the means. Line graphs and bar graphs work best. We have produced a line graph that represents the data presented in Table. This graph is presented in Figure.

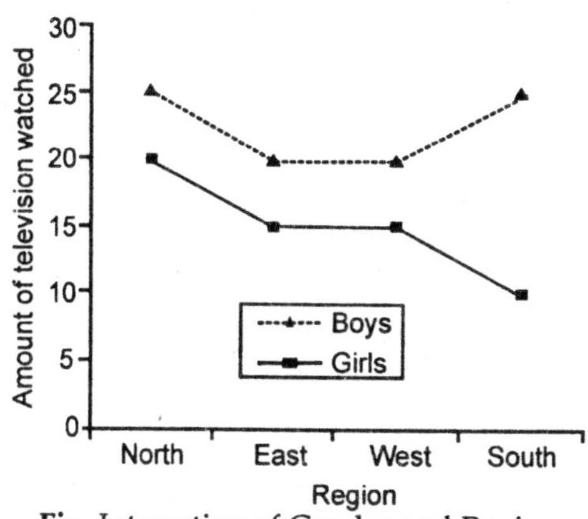

Fig. Interaction of Gender and Region

When we look at this graph, the nature of the interaction becomes readily apparent. Specifically, what we can see is that there is a consistent pattern for the relationship between gender and amount of television viewed in three of the regions (North, East, and West), but in the fourth region (South) the pattern changes somewhat. Specifically, the gap between boys and girls in the average amount of television watched per week is much wider in the South than in the other three regions. In Figure, we can see that the means for boys and girls on the dependent variable are further apart in one region than in the other, but the lines never cross. That is, there is no region *in* which girls have higher average scores than boys on the dependent variable.

As you look at the graph presented in Figure, notice that you can see both the main effects and the interaction effects. Recall that the main effect for gender indicates that, when we combine the scores from all four regions, boys appear to have higher average scores than girls on our dependent variable (*i.e.,* amount of television watched per week).

In Figure this effect is clear, as we can see that the line for boys is higher than the line for girls in all four regions. We can also see evidence of a main effect for region, although this effect is somewhat less clear than the main effect for boys. We see the region effect by noting that for both boys and girls, the average amount of television viewing is higher in the North than in either the East or West regions.

This main effect is complicated a bit, however, by the presence of the interaction. Notice that whereas the mean is lower in the South than in the North for girls (supporting our main effect for region), the mean for boys in the South is equal to the mean for boys in the North.

This raises a difficult question: When we say there is a main effect for region, with children in the North watching more television, on average, than children in the other three regions, are we being accurate?

Interpreting Main Effects in the Presence of an Interaction Effect

Researchers do not always agree on the best way to interpret main effects when there is a significant interaction effect. Some argue that it makes little sense to interpret main effects at all when there is an interaction effect present, because the interaction effect essentially modifies the meaning of the main effect.

In the preceding example, the main effect for region that shows children in the North watch more television than children elsewhere is really only true within the girls sample. In fact, boys in the South watch as much television as boys in the North, and girls in the North do not watch more television than boys in any region.

Therefore, some would argue that we should just describe the nature of the interaction, and not interpret the main effects. The logic of this argument is as follows: If we say that children in the North watch more television than children in other regions, the statement is misleading because it is not true for boys. To be accurate, we should just say that *girls* in the North watch more television that *girls* in other regions.

Others, ourself included, think it makes sense to interpret all of the effects and to consider them in relation to each other. Returning to our earlier example, we can see that there is a main effect for gender, with boys watching more television, on average, than girls.

We can also see that this effect is especially pronounced in the South. In addition, we can say that overall, when we *combine* the samples of boys and girls together, there is a main effect for region such that Northern children watch more television than children in other regions, on average.

When we add the consideration of the interaction effect, we can further argue that this overall effect is due *primarily* to differences within the sample of girls, and less to variation within the sample of boys. It is possible to get an interaction effect without a main effect.

In this example, boys and girls have equal means, as do children in each of the four geographic regions). Therefore, it makes sense to report and interpret significant main effects, even in the presence of an interaction effect. The key is to provide enough information so that readers of your results can makes sense of them. To do this, it may be necessary to discuss your interaction and main effects in relation to each other.

Here is another example to more clearly illustrate the problems of interpreting main effects in the presence of significant interactions. Suppose that we were to examine the math skills of boys and girls in two different types of mathematics programmes.

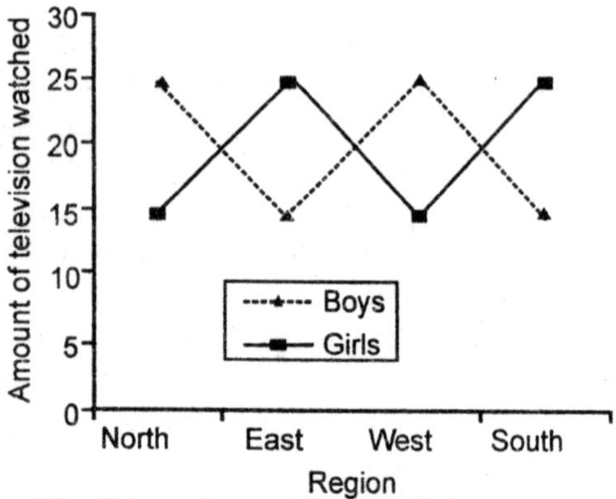

Fig. Interaction with Equal Means

Students in the "Traditional" programme study math in the usual way, reading a textbook and working out math problems in class.

Students in the "Experimental" programme work in groups to solve problems collaboratively and work with more real-world, applied problems.

After one year, we give a math test to 25 randomly selected boys and 25 randomly selected girls from each math programme. We calculate the averages for these four groups, which are presented in Figure.

The means presented in the figure clearly show that although boys and girls in the Traditional math programme had similar average scores on the math test, girls did much better than boys in the Experimental math programme. This is an interaction.

In addition, because girls in the Experimental programme did so well on their math test, the overall mean for the Experimental group was significantly higher than the overall mean for the Traditional group, thereby creating a main effect for math programme.

But does it make sense to say that students in the Experimental math programme did better on the test than students in the Traditional programme? Clearly, this was not the case for boys, and some would argue that it would be misleading to point out the main effect for math programme because the effect is only present for the girls, not the boys.

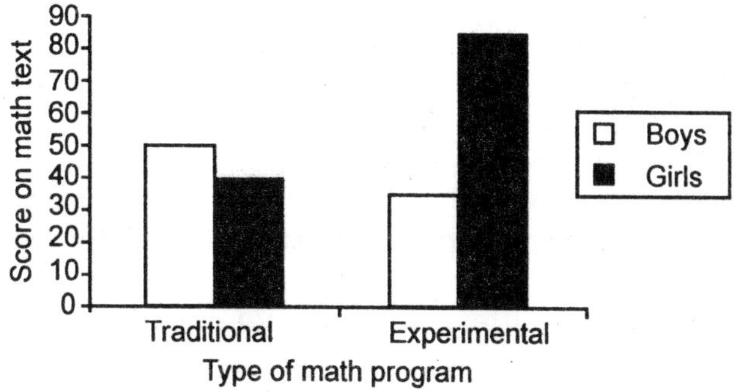

Fig. Interaction of Gender by math Programme

TESTING SIMPLE EFFECTS

Once we have found our main and interaction effects in factorial ANOVA, we can conduct one final set of analyses to examine the simple effects. The methods used to calculate the simple effects and determine whether they are statistically significant are analogous to the post-hoc tests described in Chapter.

What simple effects analysis allows us to do is to test whether there are significant differences in the average scores of any particular cells. One of the benefits of simple effects analysis is that is allows us to better understand some of the complexities in our data, particularly how to make sense of significant interaction effects.

Returning to our sample data presented in Figure we can see that we have four cells: girls in the Traditional math programme, Traditional boys, Experimental girls, and Experimental boys. With a simple effects analysis, we can test whether boys in the Traditional math programme ($X = 50$) had significantly higher average math test scores than did boys in the Experimental programme ($X - 35$). We could also test whether boys and girls in the Traditional programme differed significantly.

Perhaps most important for helping us understand the interaction effect, we can test whether girls in the Experimental programme had higher average math test scores than students in each of the three other groups.

Analysis of Covariance

Earlier in this chapter, we suggested that one of the benefits of conducting factorial ANOVAs is that it allows us to determine whether groups differ on some dependent variable while controlling for, or partialing out, the effects of other independent variables. A closely related concept that applies to all types of ANOVA, including one-way, factorial, and repeated-measures, is the use of covariates in these analyses.

In analysis of covariance (ANCOVA), the idea is to test whether there are differences between groups on a dependent variable after controlling for the effects of a different variable, or set of variables. The difference between an ANCOVA and the types of controlled variance we described earlier is that with an ANCOVA, the variable(s) that we are controlling, or partialing out, the effects of is not necessarily an independent variable. Let me explain.

In our earlier example, we was able to test whether boys and girls differed in the amount of television they watched while controlling for the effects of which region of the country they lived in (the second independent variable), as well as the interaction between the two independent variables.

But in an ANCOVA analysis, we can control the effects of variables *besides* independent variables. For example, we could use socioeconomic status (SES) as a covariate and test whether children in different regions of the country differ in the amount of television they watch *after* partialing out the effects of their SES. Suppose that our sample of children from the North is less wealthy than our samples from the three other regions. Suppose further than children from poorer families tend to watch more television than children from wealthier families.

Because of this, our earlier results that found greater television watching among children in the Northern region may simply be due to the fact that these children are less wealthy than children in the other regions. With ANCOVA, we can test whether the difference in the viewing habits of children from different regions is due strictly to differences in

SES, or whether there are regional differences *independent* of the effects of SES. This is particularly handy because even though factorial ANOVA only allows us to use categorical (*i.e.*, nominally scaled) independent variables, with ANCOVA we can also control the effects of continuous (*i.e.*, intervally scaled) variables.

Effect Size

It is illustrated effect size in factorial ANOVA, along with some of the particulars about sums of squares, mean squares, and *F* values, using output from an analysis of our own data using the SPSS computer software programme. In this example, students' confidence in their ability to understand and successfully complete their English classwork, referred to here as self-efficacy, was the dependent variable. We wanted to see whether high school boys and girls differed in their self-efficacy (*i.e.*, a main effect for gender), whether students with relatively high grade point averages (GPA) differed from those with relatively low GPAs in their self-efficacy (*i.e.*, a main effect for GPA), and whether there was an interaction between gender and GPA on self-efficacy.

Gender, of course, is a two-category independent variable. To make GPA a two-category variable, we divided students into high- and low-GPA groups by splitting the sample in two using the median GPA. This allowed me to perform a 2 (gender) x 2 (GPA) factorial ANOVA. Self-efficacy was measured using a survey with a 5-point scale (1="not at all confident" and 5="very confident"). Our sample consisted of 468 high school students.

The results presented in Table begin with descriptive statistics. These statistics are presented separately by subgroups (*e.g.*, low-achieving girls, high-achieving girls, all girls combined, low-achieving boys, etc.). The means and standard deviations presented are for the dependent variable, self-efficacy. By glancing over the means we can see that the boys in our sample had slightly higher average feelings of self-efficacy than did the girls, and this difference appears to be largest among the boys and girls in the low-GPA group.

Turning our attention to the ANOVA results, there are a number of important features to notice. In the far left column titled "Source, " there are the various *sources* of variation in self-efficacy. These are the different ways that the variance of the dependent variable, self-efficacy, is sliced up by the independent variables.

The first source is called the "Corrected Model." This is the combination of all of the main and interaction effects. If covariates were used, these effects would be included in the "Corrected Model" statistics. Reading from left to right, we can see that the full model has a sum of squares (11.402), which when divided by three degrees of freedom *("df")* produces a "Mean Square" of 3.801. When we divide this by the mean square error a few rows down (MS_e =.649) we get an F value of 5.854.

This has a "Sig." of.001 (in other words, p =.001). Because this value is less than.05, the overall model is statistically significant. But is it *practically* significant? In the final column labeled "Eta Squared" we can see that the overall model accounts for only 3.6 Per cent of the variance in self-handicapping scores. In other words, gender, GPA level, and the interaction of these two *combined* only explain 3.6 Per cent of the variance. Although this is *statistically* significant, this may not be a big enough effect size to be considered *practically* significant.

Remember that statistical significance is influenced by sample size, and 468 cases is a pretty large sample. An effect size of.036, in contrast, is not affected by sample size and therefore may be a better indicator of practical significance.

SPSS results for gender GPA factorial ANOVA

Gender	GPA	Mean	Std. derivation	N
Girl	1.00	3.6667	.7758	121
	2.00	4.0050	.7599	133
	Total	**3.8438**	**.7845**	**254**
Boy	1.00	3.9309	.8494	111
	2.00	4.0031	.8503	214

	Total	**4.0031**	**.8503**	**214**
Total	1.00	3.7931	.8208	232
	2.00	4.0381	.7989	236
	Total	**3.9167**	**.8182**	**468.**

Table. ANOVA Results

Source	**Type III**	**off**	**Mean Sum of**	**F**	**Sig. Square**	**Etc. Squard**
Corrected Model	11.402	3	6.801	5.854	.001	.036
Intercept	7129.435	1	7129.435	10981.566	.000	.959
Cendet	3.354	1	3.354	5.166	.003	.011
GPA	6.912	1	6.912	10.646	.001	.022
Gender+ GPA	1.028	1	1.028	1.584	.209	.003
Error	301.237	464	.649			
Total	**7491.889**	**468**		**Total**	**312.639 467**	

In addition to the *F* value, *p* value, and effect size for the entire model, SPSS prints out statistics for each of the main effects and the interaction effect as well. Here we can see that gender is a statistically significant predictor of self-efficacy ("Sig."=.023), but the effect size of.011 is very small. By looking at the overall means for girls and boys in the top portion of Table, we can see that boys (X_{boys} = 4.0031) have the slightly higher average feelings of self-efficacy in English than girls (*Xgirls* = 3.8438).

GPA has a larger *F* value (*F* = 10.646) and is statistically significant, but it also has a small eta squared value (.022). Students in the high-GPA group had slightly higher average feelings of self-efficacy (X_{high} = 4.0381) than did students in the low-GPA group ($X_{lo\ w}$ = 3.793). The gender by GPA interaction was not statistically significant and has a tiny effect size. Overall, then, the statistics presented in Table reveal that although there are statistically significant main effects for gender and GPA on self-efficacy, and the overall model is statistically significant, these effect sizes are quite small and suggest that there is not a strong association between either gender nor GPA with self-efficacy among this sample of high school students.

There are two other features of the SPSS output presented in Table worth noting. First, the sum of squares that SPSS uses by default in a factorial ANOVA is called "Type III" sum of squares. This means that when SPSS calculates the sum of squares for a particular effect, it does so by accounting for the other effects in the model. So when the sum of squares for the gender effect is calculated, for example, the effect of GPA and the gender by GPA interaction effects have been partialed out already. This allows us to determine the *unique* effect of each main effect and interaction effect. Second, notice that the *F* value for each effect is obtained by dividing the mean square for that effect by the mean square error. This is the same way *F* values were calculated in one-way ANOVA discussed.

REPEATED-MEASURES

One type of *t* test discussed the paired *t* test. One type of study in which a paired *t* test would be used is when we have two scores for a single group on a single measure. For example, if we had a group of third-grade students and we gave them a test of their math abilities at the beginning of the school year and again at the end of the school year, we would have one group (third graders) with two scores on one measure (the math test). In this situation, we could also use a repeated-measures analysis of variance (ANOVA) to test whether students scores on the math test were different at the beginning and end of the academic year.

Repeated-measures ANOVA has a number of advantages over paired *t* tests, however. First, with repeated-measures ANOVA, we can examine differences on a dependent variable that has been measured at more than two time points, whereas with an independent *t* test we can only compare scores on a dependent variable from two time points. Second, as discussed on factorial ANOVA, with a repeated-measures ANOVA we can control the for effects of one or more covariates, thereby conducting a repeated-measures analysis of covariance (ANCOVA). Third, in a repeated-measures ANOVA, we can also include one or more independent categorical, or group variables. This type of mixed model is a particularly useful technique and is discussed in some detail.

WHEN TO USE EACH TYPE OF REPEATED-MEASURES TECHNIQUE

The most basic form of a repeated-measures ANOVA occurs when there is a single group (*e.g.*, third graders) with two scores (*e.g.*, beginning of the year, end of the year) on a single dependent variable (*e.g.*, a mathematics test). This is a very common model that is often used in simple laboratory experiments. For example, suppose we wanted to know whether drinking alcohol affects the reaction time of adults when driving.

We could take a group of 50 adults and test their reaction time when driving by flashing a red light at each one of them when they are driving and measuring how long it takes for each one to apply the brakes. After calculating the average amount of time it takes this group to apply the brakes when sobre, we could then ask each member of our group to consume two alcoholic drinks and then again test their reaction time when driving, using the same methods. In this example, I've got one group (50 adults) with two scores on one dependent variable (reaction time when driving). After the second measure of reaction time, we could ask each of our participants to consume two more alcoholic drinks and again test their reaction time. Now we've got three measures of reaction time that we can use in our repeated-measures ANOVA. Notice that our dependent variable is always the same measure (reaction time), and our group is always the same (sample of 50 adults). The results of our repeated-measures ANOVA will tell me whether, on average, there are differences in reaction time across our three trials. If there are, we might logically conclude that drinking alcohol affects reaction time, although there may be other explanations for our results (*e.g.*, our participants may be getting tired or bored with the experiment, they may be getting used to the test situation, etc.).

In a slightly more advanced form of the test of reaction time, we could include a covariate. In the previous example, suppose that we found the reaction time was fastest when our participants were sobre, a bit slower after two drinks, and a lot slower after four drinks. Suppose that we publish these results and the national beer, wine, and liquor companies become worried that, because of our study, people will stop drinking their products for fear of getting in automobile accidents.

These producers of alcoholic drinks begin to criticize our study. They suggest that because equal amounts of alcohol generally have greater effects on those who weigh less than on heavier people, our results may have been skewed by the effects of alcohol on the lighter people in our study. "Although the effects of two alcoholic drinks may impair the reaction time of lighter people, even four alcoholic drinks will not impair the reaction time of heavier people" said the United Alcohol Makers of America (a fictitious group).

Stung by the criticism of the UAMA, we decide to replicate our study, but this time we use weight as a covariate. Again, we measure participants' reaction time when driving completely sobre, after two alcoholic drinks, and after four alcoholic drinks. In addition, this time we weigh each of our participants. Now when we analyse our data, we include our weight covariate. We find that, after controlling, or partialing out the effects of weight, there is no difference in the reaction time of participants before they have any drinks and after they have two drinks, but after four drinks our participants react more slowly, on average, than they did after zero or two drinks. These results suggest that drinking may increase the reaction time of lighter people after only two drinks, but it seems to impair the reaction time of people, regardless of weight, after four drinks.

Still bothered by our results, the UAMA suggests that our results are skewed because we did not look at the effects of drinking on reaction time separately for men and women. "Women are more dramatically affected by alcohol than men, regardless of weight" claims the UAMA. They argue that although consuming four alcoholic drinks may slow the reaction time of women, it will not have an effect on heavy men. Though we are dubious of the argument that heavy men should have their rights to drink and drive protected, in the name of science we decide to conduct one final study. In this study, again with 50 adults (25 women and 25 men) of various weights, we again test their reaction time while driving after zero, two, and four alcoholic drinks. Now I've got one dependent variable (reaction time) measured at three time points, one covariate (weight), and one independent group variable (gender of participant). Notice that although number of drinks is technically an independent variable, it is not a categorical, or group, variable. In other words, we do not have three independent groups (the zero-drink group, the two-drink group, and the four-drink group). Rather, we have three *dependent*, or repeated, measures of the same dependent variable, reaction time.

When we examine the results of our study, we find that, after controlling for the effects of our covariate (weight), there is still no difference in reaction time measured after zero and two drinks, but still slower reaction time, on average, after four drinks. In addition, we find no interaction between gender and number of drinks on reaction time. This tells us that *both* men and women have slower reaction time after four drinks, regardless of their weight.

To summarize, our three different repeated-measures ANOVAs produced the following results. The first one found that adults' reaction times while driving were slower, on average, after two drinks and slower still after four drinks. Our second test included the covariate of weight, and we found that when we control for the effects of weight, reaction time is not slower after two drinks but is slower after four drinks. Finally, in our third analysis, we examined whether changes in reaction time after two and four drinks, when controlling for weight, was different for men and women. found that it was not. These three analysis provide a snapshot of how repeated-measures ANOVA works and what information it can provide. Now let's take a closer look at how it works.

REPEATED-MEASURES ANOVA

Repeated-measures ANOVA is governed by the same general principles as all ANOVA techniques. As with one-way ANOVA and factorial ANOVA, in repeated-measures ANOVA we are concerned with dividing up the variance in the dependent variable. Recall that in a one-way ANOVA, we separated the total variance in the dependent variable into two parts: that attributable to differences between the groups, and that attributable to differences among individuals in the same group (a.k.a., the error variance). In a repeated-measures ANOVA with no independent group variable, we are still interested in the error variance. However, we also want to find out how much of the total variance can be attributed to time, or trial. That is, how much of the total variance in the dependent variable is attributable to differences *within individuals* across the tunes they were measured on the dependent variable. Consider an example. Suppose that we are interested in examining whether a group of students increase their knowledge and skills from one academic year to the next. To do this, we give our sample a standardized test of vocabulary (with a possible range of 1"100), once when they are finishing third grade and again when they are finishing fourth grade. When we do this, suppose we get the data presented in Table.

Table. Vocabulary test Scores at Two Time Points

Case Number	Text Scene, Time I (Third Grade)	Text Scene, Time 2 (Fourth Grade)
1	40	60
2	55	55
3	60	70
4	40	45
5	75	70
6	80	85
7	65	75
8	40	60
9	20	35
10	45	60

For each of the 10 cases in Table, we have two test scores, giving us a total of 20 scores in the table. We could find an average for these 20 scores, and a standard deviation, and a variance. In a repeated-measures ANOVA, we want to try to partition this total variance into different pieces. In the most basic form of repeated-measures ANOVA, there are three ways that we can slice up this variance.

First, there is the portion of variance attributable to deviations in scores between the individual cases in the sample. For each case in our sample, we have two scores (one for Time 1 and one for Time 2). We can find an average of these two scores, for each individual, and then see how much this individual average differs from the overall average. In Table, for example, the first case has an average score of 50 across the two trials (40 + 60 2=50). The overall average for the scores in the table if 56.75. So there is some variation in the average scores of the 10 individuals in the sample. This is one source of variation.

The second source of variation in the scores involves the within-subject variance, or differences, between Time 1 and Time 2 scores. As we can see by looking at the scores in Table and in Figure, it appears that students generally had different scores on the test at Time 1 than they did at Time 2. These infra-individual, or within-subject, differences between Time 1 and Time 2 scores can be seen more easily in the graph presented in Figure.

These intra-individual changes reflect differences, or variance, *within each individual,* and therefore are called within-subject effects. What we are interested in is whether, on average, individuals' scores were different at Time 1 (in third grade) than they were at Time 2 (in fourth grade). Notice that we are asking whether there were differences in the scores between Time 1 and Time 2 *on average.*

If the scores of some of the cases went up from Time 1 to Time 2, but the scores of other cases went down by the same amount, then these changes would cancel each other out, and there would be no *average* difference between the Time 1 and Time 2 scores. But if the scores either went up or down on average between Time 1 and Time 2, then we could say that some of the total variation can be attributable to within-subject differences across time. A look at the scores in Table and Figure reveals that scores *appeared* to increase from Time 1 to Time 2. To examine whether there are differences in the average scores across time, all we need to do is calculate the average score at each time and find the difference between these average scores and the overall average.

In the preceding paragraph, we found that the overall average score was 56.75. In Table, we can see that the average for Time 1 is 51.5, and the average score for Time 2 is 61.5. So we can see that there is some variance in the average scores at the two times (*i.e.,* third and fourth grade), suggesting that there may be a within-subjects effect.

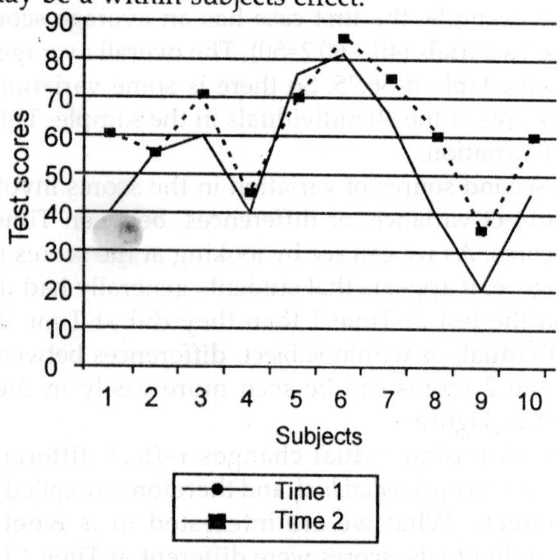

Fig. Time 1 and Time 2 test scores.

The third source of variation in the scores comes from the interaction between the within-subject scores and the variance in scores across the subjects. Although it appears that the scores of the members in our study increased, *on average,* from Time 1 to Time 2, it does not appear that these within-subject changes over time were the same across all of the subjects in the sample. As Figure clearly shows, some subjects had large increases in their test scores from Time 1 to Time 2 (*e.g.*, Subjects 1, 8, 9, and 10), whereas others had more modest increases, one had no change (Subject 2) and one actually had a lower score at Time 2 (Subject 5). So there appears to be a case, or subject, by time interaction. In other words, the size of the increase in test score from third to fourth grade depended on the which subject we were looking at. This difference among the subjects in the magnitude of change from Time 1 to Time 2 represents the third source of variance.

Using these three sources of variance, we can then calculate an *F* ratio and determine whether there are statistically significant differences in the average scores at Time 1 and the average scores at Time 2. To do this, we divide the mean square for the differences between the trials, or time, averages *(MS T)* by the mean square for the subject by trial interaction *(MS S x* r).

The degrees of freedom *F* ratio is the number of trials minus 1 *(T-* 1) and *(T – l)(S-* 1), where *S* represents the number of subjects in the sample. What we get when we calculate this *F* ratio is an answer to the following question: How large is the difference between the average scores at Time 1 and Time 2 relative to (*i.e.*, ivided by) the average amount of variation among subjects in their change from Time 1 to Time 2? Because differences in the rate of change across time are just considered random fluctuations among individuals, this *F* ratio, like all *F* ratios, is a measure of *systematic* variance in scores divided by *random* variance in scores

In this most basic form of repeated-measures ANOVA, notice that what we are primarily concerned with is whether there is a systematic pattern of differences *within individuals, or subjects,* in the scores on the dependent variable measured at two time points. Also notice that if we had three points of data (*e.g.*, test scores from third, fourth, and fifth grades), our question would remain the same: Is there a pattern of differences in the scores *within subjects* over time? Keep in mind that when we say "a pattern" or "a systematic pattern, " we mean *on average*. So, to rephrase the question, a simple repeated-measures ANOVA can help us detect whether, *on average,* scores differ *within subjects* across multiple points of data collection on the dependent variable. This type of simple repeated-measures ANOVA is sometimes referred to as a within-subjects design.

REPEATED-MEASURES ANALYSIS OF COVARIANCE (ANCOVA)

A slightly more complicated form of repeated-measures ANOVA can be produced by adding one or more covariates to the model. As well as in factorial ANOVA, covariates can be used to partial out some portion of the variance in the dependent variable. We illustrate how this works by returning to the example data presented in Table.

One could argue that the results of our repeated-measures ANOVA were skewed by the scores of the more intelligent students in our sample. Although the students in our sample scored higher on the vocabulary test at the end of fourth grade than they did at the end of third grade, we must keep in mind that the change in scores over time represents an *average* change. Some students in our sample improved quite a bit over time, whereas others did not increase at all, and one actually declined.

So it is possible that this overall average improvement over time was caused by large increases among the brightest students. To explore this hypothesis, we conduct a new repeated-measures ANOVA, but this time we include a covariate: IQ test scores.

When we conduct our repeated-measures ANOVA, we now have three ways of partitioning the total variance in our vocabulary test scores.

First, there is the portion of variance that is accounted for by our covariate, IQ test scores. If students' IQ test scores are related to (*i.e.,* correlated with) their vocabulary test scores, then the IQ test scores will explain, or account for, some percentage of the variance in students' vocabulary test scores.

Second, after partialing out the portion of variance attributable to IQ test scores, we can see whether any of the remaining variance in vocabulary test scores is accounted for by *changes* in vocabulary test scores from third to fourth grade.

In other words, once we control for the effects of IQ test scores, do the scores of our sample change significantly from Time 1 (third grade) to Time 2 (fourth grade), on average? Is there still a *within-subjects* effect after controlling for IQ test scores?

Finally, after accounting for the variance in vocabulary test scores that is attributable to the covariate (*i.e.*, IQ test scores) and the within-subjects effect (*i.e.*, changes from third to fourth grade), there will still be some variance in vocabulary test scores that is not explained.

This is error variance, which is the same as the random variance that we normally find between different members of the same sample.

To reiterate, when a covariate (or several covariates) are added to the repeated-measures ANOVA model, they are simply included to "soak up" a portion of the variance in the dependent variable.

Then, we can see whether there are any within-subject differences in the scores on the dependent variable, when *controlling for, or partialing out* that portion of the variance accounted for by the covariate(s).

In the example we have been using, the addition of the IQ score covariate allows us to answer this question: Do students' vocabulary test scores change, on average, from third to fourth grade *independently* of their IQ scores? Phrased another way, we can ask whether, when controlling for IQ, students' vocabulary test scores change from third to fourth grade.

ADDING AN INDEPENDENT GROUP VARIABLE

Now that we have complicated matters a bit by adding a covariate to the model, let's finish the job by adding an independent categorical, or *group* variable. Suppose, for example, that our 10 cases listed in Table included an equal number of boys and girls.

This two-level independent variable may allow us to divvy up the variance in our dependent variable even more, but only if there are differences in the scores of boys and girls.

Table. Vocabulary Text Scores at the Time Point

Case Number	Text Source, Time I (Third Grade)	Text Source, Time 2 (Fourth Grade)
	Boys	
9	20	35
4	30	45
8	40	60
1	40	60
10	45	60
	Girls	
2	55	55
3	60	70
7	65	75
5	75	70
6	80	85

There are two ways that this independent group variable may explain variance in vocabulary test scores. First, boys and girls may simply differ in their average scores on the vocabulary test. Suppose that when we divide the scores in Table by gender, we get the results presented in Table. If the data were aligned in this way, we would find that at Time 1, the average score on the vocabulary test was 35 for boys and 65 for girls.

Similarly, at Time 2, the average score for boys was 52 whereas for girls it was 71. At both Time 1 and Time 2, boys appear to have lower average scores on the vocabulary test than girls. Therefore, there appears to be a *main effect* for gender. Because this main effect represents a difference between groups of cases in the study, this type of effect is called a between-groups or between-subjects main effect. In other words, some of the variance in vocabulary test scores can be explained by knowing the group (*i.e.*, gender) to which the student belongs.

The second way that our independent group variable can explain some of the variance in our dependent variable is through an interaction effect. If we were to graph the means for boys and girls at both time points, we would get an interesting picture. The main effect for gender is clear. In addition, it is also clear that there is a within-subject effect, because both boys and girls have higher scores at Time 2 than they did at Time 1. But what also becomes clear in Figure is that the *amount* of change from Time 1 to Time 2 appears to be greater for boys than for girls. Whereas the average score for girls increased 6 points from third to fourth grade, it grew 17 points for boys. These different amounts of change represent another source of explained variance in vocabulary test scores: the interaction of the within-subjects effect with the *between-subjects* effect. In other words, there appears to be a gender (*i.e.*, between-subjects) by time (*i.e.*, within-subjects) interaction on vocabulary test scores.

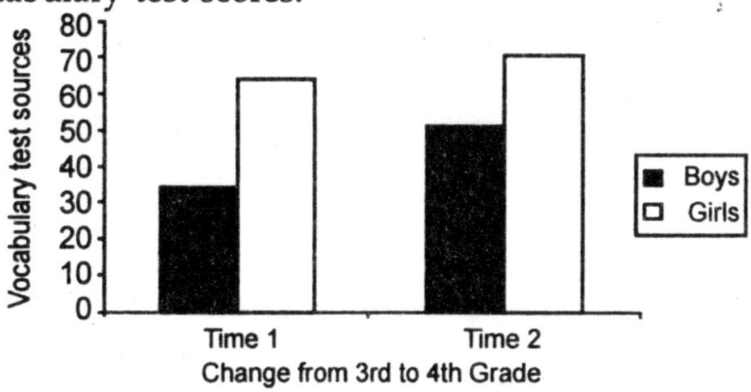

Fig. Gender by Time Interaction

To summarize, our final model has a number of effects, each of which can explain some of the variance in the vocabulary test scores of the cases in our sample. First, some of the variance in vocabulary test scores can be explained by students' IQ test scores.

On average, students with higher IQ test scores had higher vocabulary test scores. Second, even after controlling for IQ, there was a within-subjects main effect. That is, we can know something about students' scores on the vocabulary test by knowing whether we are talking about Time 1 or Time 2 test scores because, on average, students had higher scores at Time 2. Third, there was a between-subjects effect for gender, so we can explain some of the variance in vocabulary test scores by knowing the gender of the student. Girls had higher scores, on average, than did boys. Fourth, our time by gender interaction explains some additional variance in vocabulary test scores. Although both boys' and girls' scores improved over time, this improvement was more dramatic among boys, on average. Finally, there is some variance in vocabulary test scores that we cannot explain with our covariate, time, gender, or interaction effects. This is error variance.

Please keep in mind that our different effects (time, gender, interaction, covariate) will only explain variation in our dependent variable if the relations between our dependent variable and these effects are statistically significant. With only 10 cases in our sample, many of these effects may not be statistically significant.

10

The Histogram

A histogram is a graphical method of representing a probability distribution over an interval of the real number line. First, we discuss the formal representation of a histogram and follow this up with an informal discussion. We assume that one has n data points from a particular probability distribution, $x_1.....x_n$ over an interval [a, b]. To obtain histograms, we partition the interval [a, b] into m equally sized intervals called bins. Bin j is then:

$$B_j = [a + \frac{(b-a).(j-1)}{m},,,a + \frac{(b-a).j}{m} \textit{ for } j=1,....m$$

The endpoints a and b are usually taken to be the minimum and maximum of the data set. The number of data points in bin j is n_j. Of course it will turn out that

$$\sum_1^j n_j = n$$

Essentially, a histogram is the simplest non-parametric density estimator and is the that is used by most researchers. The figure below depicts a histogram of 100 pseudo random numbers from a normal distribution with a mean of zero and standard deviation of one:

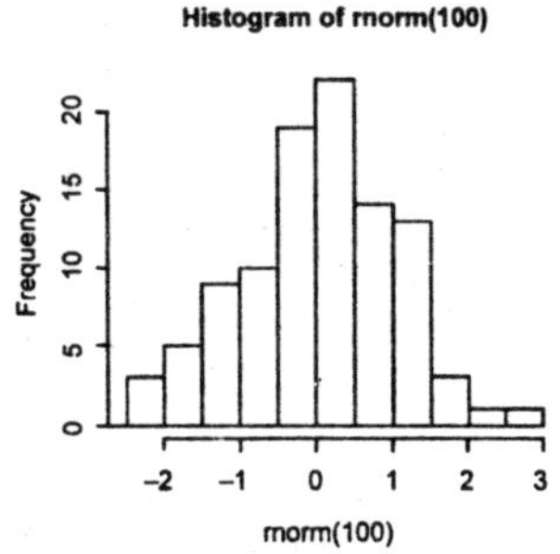

This histogram is constructed by dividing the interval covered by the data values into equal sub-intervals called bins. Each time a data point falls into a particular bin, then the bin is incremented by 1. The choice of endpoints and the choice of the number of sub-intervals can have marked effects on the shape of the histogram. Data can look bimodal when represented with a particular number of bins and bin width, but can appear uni-modal when represented with, for example, less bins and a wider bin width:

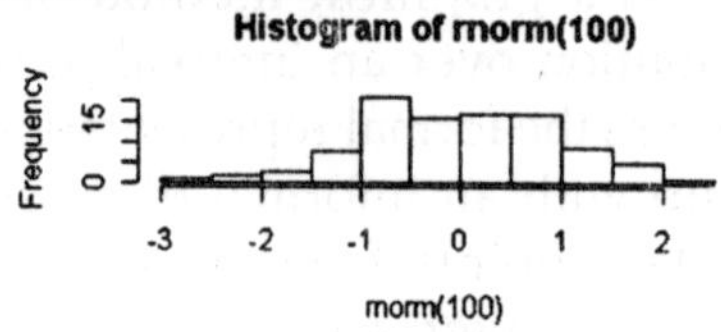

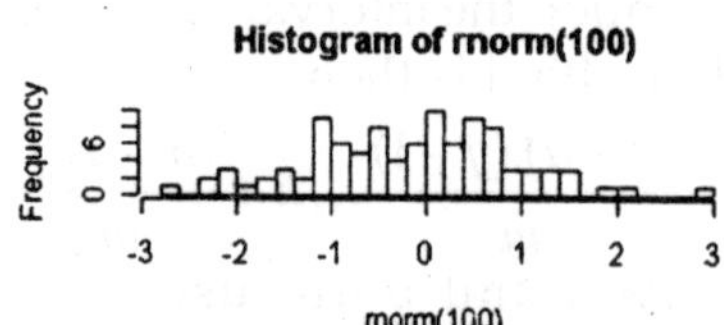

Thus, histograms have a few drawbacks:

- They are not smooth,
- The depend greatly on the end points of the bins,
- They depend on the width of the bins.

These first two problems can be addressed by a histogram smoothing technique called "kernel density estimation". Before we look at kernel density estimation, we want to illustrate a method for simulating correlated bivariate and multivariate data sets. We will use a simulated data set with a known population correlation to illustrate kernel density estimation in two dimensions.

The effect is a rough approximation of the frequency distribution of the data. The groups of data are called *classes,* and in the context of a histogram they are known as *bins,* because one can think of them as containers that accumulate data and "fill up" at a rate equal to the frequency of that data class.

Consider the exam scores of a group of students. By defining data classes each spanning an interval of 10 points and counting the number of scores in each data class, a frequency table can be constructed as in the following example:

Table: Frequency Table

Group	Count	Group	Count
0 – 9	1	10 – 19	2
20 – 29	3	30 – 39	4
40 – 49	5	50 – 59	4
60 – 69	3	70 – 79	2
80 – 89	2	90 – 99	1

To construct the histogram, groups are plotted on the x axis and their frequencies on the y axis. The following is a histogram of the data in the above frequency table.

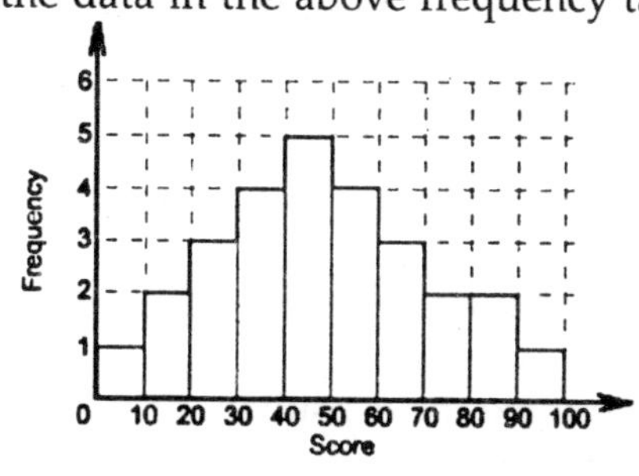

Fig. Histogram

Information Conveyed by Histograms

Histograms are useful data summaries that convey the following information:

- The general shape of the frequency distribution (normal, chi-square, etc.)
- Symmetry of the distribution and whether it is skewed
- Modality-unimodal, bimodal, or multimodal

The histogram of the frequency distribution can be converted to a probability distribution by dividing the tally in each group by the total number of data points to give the relative frequency.

The shape of the distribution conveys important information such as the probability distribution of the data. In cases in which the distribution is known, a histogram that does not fit the distribution may provide clues about a process and measurement problem.

For example, a histogram that shows a higher than normal frequency in bins near one end and then a sharp drop-off may indicate that the observer is "helping" the results by classifying extreme data in the less extreme group.

BIN WIDTH

The shape of the histogram sometimes is particularly sensitive to the number of bins. If the bins are too wide, important information might get omitted. For example, the data may be bimodal but this characteristic may not be evident if the bins are too wide.

On the other hand, if the bins are too narrow, what may appear to be meaningful information really may be due to random variations that show up because of the small number of data points in a bin.

To determine whether the bin width is set to an appropriate size, different bin widths should be used and the results compared to determine the sensitivity of the histogram shape with respect to bin size.

Bin widths typically are selected so that there are between 5 and 20 groups of data, but the appropriate number depends on the situation.

HISTOGRAMS AND BOXPLOTS

The histogram provides a graphical summary of the shape of the data's distribution.

It often is used in combination with other statistical summaries such as the boxplot, which conveys the median, quartiles, and range of the data.

Purpose of A Histogram

A histogram is used to graphically summarize and display the distribution of a process data set.

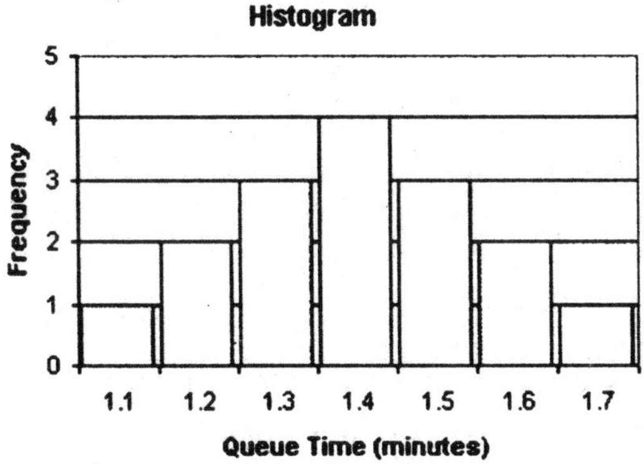

Fig. Sample Bar Chart Depiction

How To Construct A Histogram

A histogram can be constructed by segmenting the range of the data into equal sized bins (also called segments, groups or classes). For example, if your data ranges from 1.1 to 1.8, you could have equal bins of 0.1 consisting of 1 to 1.1, 1.2 to 1.3, 1.3 to 1.4, and so on. The vertical axis of the histogram is labeled Frequency (the number of counts for each bin), and the horizontal axis of the histogram is labeled with the range of your response variable. You then determine the number of data points that reside within each bin and construct the histogram. The bins size can be defined by the user, by some common rule, or by software methods (such as Minitab).

What Questions The Histogram Answers

- What is the most common system response?
- What distribution (centre, variation and shape) does the data have?
- Does the data look symmetric or is it skewed to the left or right?
- Does the data contain outliers?

PROBABILITY

A Short History of Probability

"A gambler's dispute in 1654 led to the creation of a

mathematical theory of probability by two famous French mathematicians, Blaise Pascal and Pierre de Fermat. Antoine Gombaud, Chevalier de Méré, a French nobleman with an interest in gaming and gambling questions, called Pascal's attention to an apparent contradiction concerning a popular dice game. The game consisted in throwing a pair of dice 24 times; the problem was to decide whether or not to bet even money on the occurrence of at least one "double six" during the 24 throws.

A seemingly well-established gambling rule led de Méré to believe that betting on a double six in 24 throws would be profitable, but his own calculations indicated just the opposite.

This problem and others posed by de Méré led to an exchange of letters between Pascal and Fermat in which the fundamental principles of probability theory were formulated for the first time. Although a few special problems on games of chance had been solved by some Italian mathematicians in the 15th and 16th centuries, no general theory was developed before this famous correspondence.

The Dutch scientist Christian Huygens, a teacher of Leibniz, learned of this correspondence and shortly thereafter published the first book on probability; entitled *De Ratiociniis in Ludo Aleae,* it was a treatise on problems associated with gambling. Because of the inherent appeal of games of chance, probability theory soon became popular, and the subject developed rapidly during the 18th century. The major contributors during this period were Jakob Bernoulli and Abraham de Moivre.

In 1812 Pierre de Laplace introduced a host of new ideas and mathematical techniques in his book, *Théorie Analytique des Probabilités.* Before Laplace, probability theory was solely concerned with developing a mathematical analysis of games of chance. Laplace applied probabilistic ideas to many scientific and practical problems. The theory of errors, actuarial mathematics, and statistical mechanics are examples of some of the important applications of probability theory developed in the 19th century. Like so many other branches of mathematics, the development of probability theory has been stimulated by the variety of its applications. Conversely, each advance in the theory has enlarged the scope of its influence. Mathematical statistics is one important branch of applied probability; other applications occur in such widely different fields as genetics, psychology, economics, and engineering. Many workers have contributed to the theory since Laplace's time; among the most important are Chebyshev, Markov, von Mises, and Kolmogorov.

One of the difficulties in developing a mathematical theory of probability has been to arrive at a definition of probability that is precise enough for use in mathematics, yet comprehensive enough to be applicable to a wide range of phenomena. The search for a widely acceptable definition took nearly three centuries and was marked by much controversy. The matter was finally resolved in the 20th century by treating probability theory on an axiomatic basis. In 1933 a monograph by a Russian mathematician A. Kolmogorov outlined an axiomatic approach that forms the basis for the modern theory. Since then the ideas have been refined somewhat and probability theory is now part of a more general discipline known as measure theory."

Introductory Note

The term Probability, as the name suggests, refers to a calculation, a mathematical estimate concerning the possibility of a particular outcome in a situation where there are a number of different possible outcomes. If, for example, we know that in a certain situation (*e.g.*, a roll of the dice) there is a determined number of possible outcomes (*e.g.*, six possible faces of the die), we can calculate the probability of a particular outcome or of a particular sequence of outcomes if the event is repeated.

For example, with a normal die in a fair roll, there is a 1 in 6 chance of any particular face of the die showing on the top (since there are six surfaces on the cube); in a fair toss of a normal coin there is a 1 in 2 chance of a head appearing on top, a 1 in 2 chance of a tail appearing on top, and a 2 in 2 chance of a head or a tail appearing on top.

It is important to stress at the outset that probability is a *theoretical* calculation of the likelihood of a particular outcome. Thus, it is based, not upon observation of actual events (*e.g.*, listing the results from several tosses of the coin) but on calculations which are based on certain assumptions (*e.g.*, that the coin or die toss is fair, that there are no hidden factors, like cheating, influencing the outcome).

Because the probability of an outcome is a theoretical calculation, knowing the probability does not mean that we can always predict a particular outcome with certainty (please understand this very important point). Often the result of a particular event or series of events will not match the theoretical probabilities. For example, in a sequence of ten coin tosses, theory tells us that the most probable result will be 5 outcomes with heads on top and 5 outcomes with tails on top. But that calculation does not mean that, if we observe an actual 10-toss sequence, we will always get that most probable result.

Checking probabilities (that is, the theoretical calculations) against the observed outcomes of a real event is, as we shall see, an essential part of the study of probability and statistics (and a central exercise in this module). But always keep in mind this opening point: the probability of an event is a theoretical calculation; the observed frequency with which an event actually occurs is something different: the first is based upon mathematical analysis of an event (independent of observation); the second is based upon an actual experiment.

Probability theory has a number of extremely important uses. A simple but powerful application is its ability to assist us in establishing the mathematical likelihood of a particular outcome and thus to advise ourselves of the prudence of a certain action (*e.g.*, betting). For instance, to take a very simple example, in a fair toss of a coin, theory indicates that the probability of the coin landing with heads on top is 50 per cent, with tails on top 50 per cent, and with either heads of tails on top 100 per cent. In a bet on the result of a coin toss, therefore, one would be unwise to wager, say, ten dollars that a head would appear against your opponent's bet of two dollars that a tail would appear.

In the course of this module, we will become more familiar with a number of important applications of probability theory.

EXPRESSING PROBABILITY MATHEMATICALLY

We can express the probability of an event mathematically in a number of ways, but the most common is as a decimal between 1 and 0. Thus, a probability of 1 ($p = 1$) is certainty that the event will occur. A probability of 0 ($p = 0$) is certainty that event will not occur. Thus, all probabilities for an event are expressed as a decimal between 0 and 1.

The probability that a single fair coin toss will produce a head on top is expressed as follows: p =.5. This we might also express as meaning that the probability is 5 out of 10 or 50 per cent. The decimal notation is the most common and will be standard throughout the rest of this workbook.

In the same way, if p =.75, that means that the outcome will theoretically occur 75 times out of 100; if p =.234, that means that the outcome will theoretically occur 234 times out of 1000; if p =.01345, that means the outcome will theoretically occur in 1345 cases out of 100,000.

It will be clear that any value for p greater than.5 means that the result is theoretically more likely to occur than not; a value for p less than.5 means that the result is more likely not to occur than it is to occur. The smaller the value of p, the less likely that outcome.

Probability Theory and Observation

One of the major applications of probability is to check the theoretical possibility of a particular outcome (calculated mathematically) with the observed result of an event. If we find that the observed frequency of a particular occurrence departs widely from theoretical probabilities (*i.e.*, the result is occurring more frequently than we would expect from our calculations), sometimes we have good reason to investigate the event more carefully.

For example, in a fair toss of a coin, we theoretically calculate p as.5 for a head to appear on any single toss. But what is the probability that someone will throw the coin four times in a row and have the outcome heads each time? We can calculate the probability by multiplying the probabilities together. Since there are four tosses and the probability of a head is.5 on each of these, the probability of a sequence of four heads is as follows:

$$p = .5 \times .5 \times .5 \times .5 = .0625$$

In other words, in a sequence of four consecutive coin tosses, probability theory indicates that in 625 cases out of 10,000 (or 1 outcome in 16) four heads will come up in a row. Thus, if you were making a bet in a number of different trials of four consecutive coin tosses that four heads would appear in a row, you could theoretically expect to win 1 out of every 16 trials and lose 15 out of 16 trials.

Now, this figure of 1 in 16 ($p = .0625$) does not claim that in every 16 trials of a four-coin-toss sequence, one result will *always* be four heads. The theory states that, on average, in a sequence of four-toss trials, the most likely outcome is that there will be four heads appearing in 1 out of every 16 trials. But in any particular series of four-toss trials, it is quite possible for more than sixteen four-toss trials to happen before the first four-head result appears or for a four-head result to occur before sixteen trials are completed or more than once in one sixteen four-toss sequence. As we shall see, the more times one repeats the four-toss sequence, the closer the observed result will come to the theoretically calculated result.

Thus, given that one four-head sequence every sixteen throws is the most likely (but not the inevitable) outcome of a series of four-toss sequences, if you were betting on the outcome of many four-toss sequences (that there would be four consecutive heads appearing), for the game to be truly fair, you should put up $1 against your opponent's $15 for each four-coin sequence (*i.e.*, if four heads appear you win $15; if they do not your opponent wins $1). With a great many four-toss trials, you and your opponent should come out more or less equal.

Hence, if you are betting $1 that four heads will appear in a four-toss sequence of coins and if your opponent is unwilling to put up $15, you should not undertake the bet (the probabilities are against you). If she is willing to put up more than $15, then you should take the bet (the probabilities are in your favour).

An Introductory Note about Significance

When the probability of an event (theoretically calculated) is very low and yet the event occurs, sometimes we are entitled to be suspicious or alarmed at the result. Such moments are an invitation to investigate the event to see if there are hidden factors at work (*e.g.*, loaded dice, cheating, trick coins, some hitherto unknown cause). When do such moments occur? At what point is the probability so unlikely (*i.e.*, so low) that we should investigate further?

Well, let us first, as an example, consider the probability of someone's tossing a coin with ten consecutive heads appearing. Using the same way of calculating the probability as we did above (multiplying together the probabilities for a single event), we can calculate the theoretical probability of ten consecutive heads appearing as follows:

$$p = .5 \times .5 \times .5 \times .5 \times .5 \times .5 \times .5 \times .5 \times .5 \times .5 = .000977$$

This rounded off result indicates that the probability for ten consecutive heads in a series of ten fair coin tosses is 977 out of 1,000,000. Obviously the probability here is much smaller than for the four-head sequence ($p = .0625$), and we would be right to be much more surprised at such a result. But do we have the right to be suspicious? In other words, how improbable does an event have to get before we begin to think that there is something strange going on?

Scientists and social scientists have set conventional standards for how much an observed result has to differ from the theoretical expectation in order for the result to be considered significant (*i.e.*, worthy of notice or of further investigation, because it is not likely to have occurred by chance). That level is $p = .05$ (or 5 results in 100). If there is more than a probability of.05 (*i.e.*, if p is greater than.05) that an event could have occurred by chance, scientists do not usually pay much attention to it. If p is less than.05, however, then the result is considered significant and warrants analysis, because something other than chance may be affecting the outcome. Thus, for example, someone's tossing a coin and getting four consecutive heads (where p =.0625) would not prompt any enquiry; the probability is low, but not so low that we need to start wondering about the fairness of the toss (remember the cut off point is p =.05).

However, someone's tossing a coin and getting ten heads in a row (p =.000977 or, rounded off.001) is significant. The probability is so low (1 chance in 1000) that we have some grounds for thinking the result might not be produced by chance and that it is time to check the coin.

The justification for this convention of a probability of.05 as the dividing line between significant and non-significant results is not mathematical but methodological and practical. Given the constraints, economic for example, upon investigation, it has been found to pay, in terms of success in explaining the world, to concentrate research in areas where such a degree of significance is obtainable. In some tests, as we shall see, where the procedure requires a much higher level of certainty, the cut-off line is sometimes reduced to.01 (or 1 chance in 100). In general, however, the figure of.05 probability is the accepted level below which results become significant and above which we can still ascribe the results to chance.

Note the very important point that a low probability (*e.g.*, p =.001) does not indicate impossibility. It may well be that the ten-head coin-toss sequence, although very very improbable, is fair (*i.e.*, produced by chance). The very low figure for p is a warning that something may be (and probably is) interfering with the result, but the figure does not *prove* such interference.

These basic principles indicate one important aspect of probability, its use as a diagnostic tool, a means of recognizing that in a natural event something may be interfering with the results (*e.g.*, cheating or some hidden factor). Significant discrepancies between theoretical calculations of probable outcomes and the observed results of actual outcomes are often the first clue that something needs investigating. And from this initial observation, the investigators often discover something they had not known about before (*e.g.*, a new disease, like Legionnaire's Disease or AIDS, or a new natural phenomenon).

The Fundamental Basis of Probability Theory

The basis upon which the mathematical treatment of probability rests is as follows: we assume that the phenomenon we wish to investigate can be described in terms of a given number of different possible but equally probable outcomes. For instance, in a simple coin toss there are two equally probable outcomes, heads or tails.

In one roll of a die, there are six equally probable outcomes, represented by the six sides of the die, each with a number of dots from 1 to 6. In drawing a card from a full and well shuffled normal deck with no jokers, we know that there are four equally probable outcomes for the suit of the drawn card (hearts, diamonds, spades, or clubs), there are two possible outcomes for the colour of the card (red or black), and there are thirteen possible outcomes for the value of the card. Naturally we are assuming that in all these trials the procedure is fair (*i.e.*, no cheating).

Multiple outcomes (*e.g.*, so many heads in consecutive coin tosses or so many sixes in consecutive rolls of a single dice) can be calculated from the probability of a single event (as we did above by multiplying the frequency of repeated events by the probability of the single outcome in one event). Events which cannot be described in terms of equally probable outcomes are beyond the scope of probability theory discussed here.

The probability of any specific outcome is given by the number which is equal to the number of possibilities which involve that outcome divided by the total number of possibilities. For example, in a simple coin toss, there is only one possibility that a head will appear, and the total number of possible outcomes is two. Thus, the probability that a head will appear is 1 divided by 2 or.5.

In a roll of a single die, there is one chance that a six will appear, and there are six possible outcomes. Therefore the probability of a six appearing on a single roll of a normal die is 1 divided by 6 (p =.1667). If we draw a card from a normal full deck, there is only one chance that a heart will be on the card, and there are four possibilities for different suits. Therefore the probability of drawing a heart from a full deck is 1 divided by 4 (p =.25).

What is the probability that on a single roll of a die, the result will be an even number? Well, there are three possible even outcomes (the numbers 2, 4, and 6) and there are 6 possible outcomes (1, 2, 3, 4, 5, 6). Thus, the probability that on a single roll of a die the result will be an even number of dots is 3 divided by 6: p =.5.

More complicated outcomes (*e.g.*, four heads in a row in consecutive coin tosses, three sixes in a row in consecutive rolls of the die, four number 12 results in four consecutive spins of the roulette wheel, and so on) are more complicated, but we can work out the possible outcomes for such events and calculate the probabilities involved (as we did above for sequences of four and ten coin tosses).

Randomness

Random phenomenon...outcome cannot be predicted, but that has a regulardistribution over many repetitions. Probability...proportion of times an event occurs in many repeatedevents of a random phenomenon. Independent events (trails)...outcome of an event (trial) does not influence theoutcome of any other event (trial). Example of independent events: Rollinga single die twice. The result on the second roll is independent ofthe first. Example of events not independent: Picking twocards from a deck, one at a time, without replacement. (If you pick acard, and don't put it back in the deck and reshuffle, then theprobability of a red card on the second pick is dependent upon whatthe first card happened to be.)

Probability Models

- *Overview*: It is often important andnecessary to provide a mathematical description or model forrandomness. This section provides insights into how this can bedone.

- *Sample space*: Theset of all possible outcomes of a random phenomenon.
- *Example*: Consider the sum obtained when two dice are rolled.

The following table displays the sums that can be obtained from thisevent.

		RED DIE					
		1	2	3	4	5	6
GREEN	1	2	3	4	5	6	7
DIE	2	3	4	5	6	7	8
	3	4	5	6	7	8	9
	4	5	6	7	8	9	10
	5	6	7	8	9	10	11
	6	7	8	9	10	11	12

The sample space contains 11 outcomes: S = {2,3,4,5,6, 7,8,9, 10,11,12}

Probability Rules

- If A is an event, then the probability P(A) is anumber between 0 and 1, inclusive.
- Prob(A does not happen) is 1-P(A).
- Two events are disjoint if they have nooutcomes in common. If A and B are disjoint, then P(A or B) = P(A) +P(B). Some sources use the phrase mutuallyexclusive to describe events that aredisjoint.
  - In a roll of two dice, if A = "roll asum of 7" and B = "roll a sum of 2", then P(A or B) = 6/36 + 1/ 36 =7/36, since A and B are disjoint.
  - If C = "roll a sum of 7" and D = "roll an oddsum" then C and D are not disjoint since 7 is an odd sum. In thiscase, P(C or D) = P(C) + P(D) – P(C and D) = 6/36 + 18/36 – 6/36 =18/36 = 1/2.

Two events are independent if knowing that oneoccurs does not change the probability of the other. If events A andB are independent, then prob(A and B) = prob(A)xprob(B).

- If A is the event "getting a head onthe first coin toss" and B is the event "getting a head on a secondtoss of the coin." then A and B are independent, and prob(A and B) =(1/2)(1/2) = 1/4.
- If C is the event "getting a red card by pickinga card from a randomly shuffled deck," and D is the event "getting ared card by picking a second card from the deck" (when the first cardhas not been replaced), then C and D are not independent. Prob(C) =26/52 = 1/2. If the first card is red, then prob(C and D) =(26/52)(25//51), which is not equal to (26/52)(26/52). If the firstcard was replaced in the deck, and the deck reshuffled before eventD, then C and D would be independent.

More About Probability

Overview: There are basic laws thatgovern wise and efficient uses of probability. An understanding ofthese laws is important for those who wish to utilize statistics inpractical and meaningful ways..

For any two events A and B.

Prob(A or B) = Prob(A) + Prob(B) -Prob(A and B).

If A and B are disjoint, then Prob(A and B) =0.

Prob(B | A) = prob(A and B)/prob(A).

Consider a family with two children. There arefour possible boy/girl combinations, all equally likely.

Oldest	Youngest
G	G
G	B
B	G
B	B

- Prob(at least one child is a girl) = 3/4.

- Prob(two girls) = 1/4.
- Prob(two girls | one child is a girl) = 1/3.Sample space is {(G,G), (G,B), (B,G)}.
- Prob(two girls | oldest child is a girl) = 1/2.Sample space is {(G,G), (G,B)}.
- Prob(exactly one boy | one child is a girl) =2/3.
- Prob(exactly one boy | youngest child is a girl) =1/2. OK, now suppose that a family unknown to you hastwo children. Here are two questions relating to this family.

...(1) If one of the children is a boy,what is the probability that the other child is aboy?

...(2) If the oldest child is a boy, whatis the probability that the other child is a boy?

Many intelligent people think that 50% is the answer to both (1) and(2). You should be able to show that this is not the case.

BINOMIAL DISTRIBUTION

The probability distribution of a random variable X which assumes integral values x=0..., n with the probabilities,

$$P\{X = x\} = b_x(n,p) = \binom{n}{x} P^x (1-p)^{n-x}$$

where p is a parameter of the binomial distribution, called the probability of a positive outcome, which can take values in the interval $0 \leq p \leq 1$. The binomial distribution is one of the fundamental probability distributions connected with a sequence of independent trials.

Let $Y_1,\ Y_{1...,}$ be a sequence of independent random variables, each one of which may assume only one of the values 1 and 0 with respective probabilities and (*i.e.* all are binomially distributed with). The values of may be treated as the results of independent trials, with if the result of the -th trial is "positive" and if it is "negative". If the total number of independent trials is fixed, such a scheme is known as Bernoulli trials, and the total number of positive results,

$$X = Y_1 + ... + Y_n, \quad n \geq 1,$$

is then binomially distributed with parameter P.

The mathematical expectation Ez^X (the generating function of the binomial distribution) for any value of is the polynomial $[pz+(1-p)]^n$, the representation of which by Newton's binomial series has the form

$$b_0+b_1z+...+b_nz^n$$

(Hence the very name "binomial distribution".) The moments (cf. Moment) of a binomial distribution are given by the formulas,

$$EX=np,$$

$$DX=E(X-np)^2=np(1-p),$$

$$E(X-np)^3=np(1-p)(1-2p)$$

The binomial distribution function is defined, for any real y,0<y<n, by the formula,

$$F(y)=\mathrm{P}\{X\le u\}=\sum_{x=0}^{[y]}\binom{n}{x}p^x(1-p)^{n-x},$$

where is the integer part of, and,

$$F(y)=\frac{1}{B([y]+1,n-[y])}\int_p^1 t^{[y]}(1-t)^{n-[y]-1}\,dt,$$

B(a, b) is Euler's beta-function, and the integral on the right-hand side is known as the incomplete beta-function.

As, $n\to\infty$ the binomial distribution function is expressed in terms of the standard normal distribution function by the asymptotic formula (the de Moivre–Laplace theorem):

$$F(y)=\Phi\left[\frac{y-np+0.5}{\sqrt{np(1-p)}}\right]+R_n(y,p),$$

where,

$$R_n(y,p)=O(n^{-1/2})$$

uniformly for all real. There also exist other, higher order, normal approximations of the binomial distribution.

If the number of independent trials *n* is large, while the probability *P* is small, the individual probabilities $b_x(n, p)$can be approximately expressed in terms of the Poisson distribution:

$$b_x(n,p)=\binom{n}{x}p^x(1-p)^{n-x}\approx\frac{(np)}{x!}e-^{np}$$

If $n \to \infty$ and $0 < c \le y \le C$ (where and are constants), the asymptotic formula,

$$F(y) = \sum_{x=0}^{[y]} \frac{\lambda^x}{x!} e^{-\lambda} + 0(n^{-2}),$$

where, is uniformly valid with respect to all in the interval.

$$\lambda = (2n - [y])p/(2 - P).$$

11

The Binomial Distribution

Many events have only two possible results: success or failure. In tossing a coin, there are only two possible results—heads and tails—each with a probability of.5. Success in a coin toss is calling the correct result. In answering multiple choice tests where there are four possible answers for each question, there are again only two possible results for each question, a correct answer or an incorrect one. If we are trusting entirely to chance in answering such a test (*i.e.*, guessing), we have a one in four chance of getting the answer for any question correct (*i.e.*, on each question for success p =.25, for failure p =.75).

Sorting out the probabilities for a particular outcome requires us to figure out all the possible outcomes and to calculate the distribution for all those possible outcomes. And these calculations will vary depending on the number of possible outcomes. Thus, there are many possible probabilities for different events (*e.g.*, the distribution of outcomes—heads or tails—for a four coin-toss sequence will be different from the distribution of outcomes—heads or tails—for a ten coin-toss sequence; the distribution for the results of a repeated event with a.5 probability will be different from a similar sequence with a.1667 probability, and so on).

Constructing such a mathematical calculation for a particular series of outcomes is called calculating the binomial distribution. Constructing a binomial distribution for a particular sequence of events is not very difficult, but it is rather time consuming. Fortunately we have tables of binomial distributions so that we do not have to go through the calculations here. However, in this section we are concerned with being above to read and use a simple binomial distribution table, so that we can judge the probabilities of simple events with only two possible outcomes (success or failure). The simple table below provides an example of two binomial distributions. An explanation of the table follows immediately after it.

Table: Selected Binomial Distributions

p =.5	n = 10			n = 20	
r	**ind**	**com**	**r**	**ind**	**com**
0	.001	1-	0	0+	1-
1	.010	.999	1	0+	1-
2	.044	.989	2	0+	1-
3	.117	.945	3	.001	1-
4	.205	.828	4	.005	.999
5	.246	.623	5	.015	.994
6	.205	.377	6	.037	.979
7	.117	.172	7	.074	.942
8	.044	.055	8	.120	.868
9	.010	.011	9	.160	.748
10	.001	.001	10	.176	.588
			11	.160	.412
			12	.120	.252
			13	.074	.132
			14	.037	.058
			15	.015	.021
			16	.005	.006
			17	.001	.001
			18	0+	0+
			19	0+	0+
			20	0+	0+

Note that the entry 1- means a probability larger than.9995 but less than 1 (about as certain as one can be); the entry 0+ means a probability less than.0005 but greater than 0 (extremely unlikely).

This table indicates two different binomial distributions. The one on the left side indicates the probabilities of success in a series of 10 coin tosses; the one on the right side indicates the probabilities of success in a series of 20 coin tosses; this distinction is indicated by the label n = 10 on the left and n = 20 on the right. For both tables the probability of success in any one toss is.5 (hence heading p =.5 below the title).

The first column on each side has the heading r. This indicates the number of successes in the sequence. So, for example, on the left side, the values go from 0 to 10, indicating the possible outcomes of 0 heads in 10 tosses of the coin to 10 consecutive heads in 10 tosses. On the right side, the r values go from 0 to 20, indicating the possible outcomes from 0 heads in 20 tosses to 20 heads in 20 tosses.

The *ind* (= individual) column in the middle of each side of the table gives the probability of one's achieving that exact r value in a full sequence of tosses. So on the left side, for example, the probability of getting exactly 5 heads in a sequence of 10 tosses is.246 (the *ind* value on the same line as the r value of 5). In the right side of the table, the probability of getting exactly 5 heads in a sequence of 20 tosses.015.

The *cum* (= cumulative) column gives the probability of r or more successes in the total number of trials (or at least r successes). So, for example, on the left side of the table, the chances of getting 6 or more (or at least 6) heads in 10 consecutive tosses of a coin is.377; in the right side of the table, the probability of getting 5 or more (or at least 5) heads in 20 tosses of the coin is.994.

Look on the right side of the table at the r value 4. As you would expect, the probability of getting exactly four heads in a 20-toss sequence (the *ind* value) is very low (.005), but the chances of getting at least 4 heads in a 20-toss sequence (the *cum* value) is very high (.999, close to certainty).

Table: Binomial Distribution for a Six-Sided Die

		p =.1667 n = 24			
r	**ind**	**cumr**	**r**	**ind**	**cum**
0	.013	1-	1	.060	.987
2	.139	.927	3	.204	.788

4	.214	.584	5	.171	.371
6	.108	.200	7	.056	.091
8	.024	.035	9	.008	.012
10	.003	.003	11	.001	.001

The other ind and cum values from r =12 to r = 24 are 0

Some Observations on Binomial Distributions

The binomial distribution, as illustrated in Tables above, is appropriate only when the events in the sequence are independent of each other (*i.e.*, when the outcome of one event does not affect the outcome of another).

For example, in any coin toss the probability of a head is always.5; it is not affected by the result of any outcome before it. In the roll of a single die, the probability of a 6 showing as the outcome is always the same (p =.1667), no matter what the results of previous rolls might have been. In other words in sequences of coin tosses or die rolls, the outcome of any single event is independent of the outcome of any other event.

This independence does not always apply in probability situations. For example, when one draws a single card from a full, well-shuffled deck, the probability of the outcome producing a red card is.5, because 26 of the 52 cards are red (26 divided by 52 equals.5).

However, if we draw a red card and do not return it to the deck, then the probability of a second draw producing a second red card is no longer.5. Since one red card has been removed from the deck on the first draw, there are now only 51 cards in the deck, 26 black cards and 25 red ones. Therefore, the probability of drawing a red card on the second draw is now 25/51 (p =.4902); whereas, the chance of drawing a black card on the second draw with one red card removed is now 26/51 (p =.5098).

In this example, the probabilities in the second draw have been affected by the result of the first draw (one card less in the deck, one red card less in the deck). What this means is that the probability of a red card in the second draw is not independent of the first draw. Thus, the sort of binomial distribution which we have been considering in the coin toss or roll of the die (where the probability remains the same in each toss or roll) would not be appropriate.

In any card game where the deck is not full and shuffled before each event, the probabilities are constantly changing as cards are played and removed from the deck. This is the case, most famously, with blackjack, where after every hand the played cards are normally collected and placed on the bottom and a new hand dealt from the remaining cards. A careful player who keeps close track of the cards which have been played and not returned to the deck and who understands probabilities very well can calculate when the probabilities of a favourable outcome shift in her favour (*i.e.*, when p for success becomes greater than.5).

This is the basis for a number of well-known and successful legal systems for beating the casino dealer, since this is the only casino game in which the probabilities occasionally favour the player.

Here, for example, is a simple illustration. Suppose in playing blackjack at the casino (in a game between you and the dealer only), you keep careful track of all the cards which have been played and put on the bottom of the deck. Suppose, as a result of this careful observation, you know that there are only five cards left unplayed and, further, that these five are 3 eights and 2 sevens. In such a situation you pawn everything you have, sell your car, bet everything, and stand pat with the two cards dealt to you. Since in routine casino blackjack the dealer cannot stand pat with 16 or under (and you know that from the remaining cards he must have 16, 15, or 14), it is certain that the dealer will have to draw and that his hand will be over 21. In other words, the probability for your success here is 1 (certainty).

That is a very simple example. Sophisticated mathematicians have developed systems of keeping track of the shifting probabilities in blackjack, betting low when the probability is less than.5 and high when the probability for success shifts to above.5. Over time, given the fairness of the game, such players will win (and some have won enormous amounts). Casinos routinely cope with such gamblers, when they recognize them, by shuffling the deck before every hand, thus restoring the favourable probabilities to the dealer in every hand. At this point, the sophisticated gambler with a system moves on down the strip to another casino.

BERNOULLI THEOREM

The (historically) original form of the (weak) law of large numbers. The theorem appeared in the fourth part of Jacob Bernoulli's book Ars conjectandi (The art of conjecturing). This part may be considered as the first serious study ever of probability theory.

The theorem deals with sequences of independent trials, in each one of which the probability of occurrence of some event ("success") is *P*. Let *n* be the number of trials and let *m* be the random variable equal to the number of successful events. The Bernoulli theorem states that, whatever the value of the positive numbers ε and η, the probability $\supset$ of the inequality,

$$-\varepsilon \leq \frac{m}{n} - p \leq \varepsilon$$

will be higher than $1-\eta$ for all sufficiently large $\eta(n \geq n_0)$. The proof of this theorem, which was given by Bernoulli and which was exclusively based on a study of the decrease of probabilities in the binomial distribution as one moves away from the most probable value, was accompanied by an inequality which made it possible to point out a certain bound for the given $n_0$if and were given. Thus, it was found by Bernoulli that if p =2/5, the probability of the inequality,

$$-\frac{1}{50} \leq \frac{m}{n} - \frac{2}{5} \leq \frac{1}{50}$$

will be higher than 0.999 if $n \geq 25{,}550$. By introducing a slight improvement in the original reasoning of Bernoulli, it is possible to conclude that it is sufficient to select a value of *n* obeying the condition,

$$n > \frac{1+\varepsilon}{\varepsilon^2} \log \frac{1}{\eta} + \frac{1}{\varepsilon},$$

which gives in turn, for the probability $1 - \mathrm{P}$ of the inequality

$$\left|\frac{m}{n} - p\right| > \varepsilon,$$

an estimate of the form,

$$2\exp\{-\frac{1}{2}n\varepsilon^2\}$$

The condition obtained for the above example is $n \geq 17,665$ (more sophisticated estimates show that it is sufficient to take $n \geq 6502$; one may note, for the sake of comparison, that the de Moivre–Laplace theorem yields 6498 as the approximate value of n_0). Other estimates for $1 - P$ may be obtained using the Bernstein inequality and its analogues.

An Investigation of Bernoulli's Theorem

The rest of this module describes an experiment to illustrate and test Bernoulli's Theorem, to see whether our observed results get closer and closer to the calculated probabilities as we increase the number of trials.

Equipment

For this experiment you need to collect together the following items:

- Ten coins, preferably all the same size and in good condition;
- A can or cup in which to shake the coins, which must be able to tumble freely in the container when you hold your hand over the opening and shake the coins;
- A tally sheet to keep track of the results (number of heads) of head throw (the tally sheet in given below)

Procedure

Find a quiet place at a table onto which you can spill the coins without bothering your family, friends, neighbours, or the cat.

Then go through the following steps:

1. Carefully shake out the ten coins so that they are well mixed and then spill them out on a table or similar surface. Count the number of heads, and enter the number in the appropriate place on the tally sheet

on the next page (using a vertical stroke in the appropriate box for each head; for example, if on the first trial you get six heads, then put a vertical line on the six line in the table, under the Trials 1-25 column; then on the second trial, if you get three heads, then put a line in the three box in the same column, and so on).

2. Collect the data for the first 25 tosses (each with 10 coins) in the first column, so that at the end of the 25 trials there is a record of how many tosses produced no heads, one head, two heads, three heads, and so on. Then do the same for the next 25 trials, entering the results in the second column. Enter the data for the third group of 25 tosses in the third column and for the fourth group of 25 trials in the fourth column. Now you have a record of the number of heads in 100 tosses of 10 coins.
3. Calculate the average number of heads for each of the 25 toss sequences, and enter that figure in the bottom row (labelled Mean) under the appropriate column. Calculate the average by adding up the numbers in the column (to get the total number of heads) and dividing by 25 (*i.e.*, the number of trials). Note that you add up the number of heads in any one square by multiplying the number of vertical strokes times the number which the square represents. For example, if you have four vertical strokes in the 7 box (indicating that on 4 occasions there were 7 heads in the 10 coin result), then the number of heads for that box is 4 x 7 or 28.

   Thus, you will have four figures for the average number of heads in each of the four 25-toss sequences.
4. Add up the totals for the first two 25-trial sequences, and enter them in the appropriate boxes in Column 5 (headed Total 1-50). Then add up the totals for the first three 25-trial sequences, and enter them in the appropriate boxes in Column 6 (headed Total 1-75).

Then do the same for all the trials, entering the results in the boxes under the heading Total 1-100.

5. Then calculate the average number of heads (the Mean) for each column, once again adding up the total number of heads and dividing by the number of trials. Thus, you should produce an average number of heads for the first 50 trials, the first 75 trials, and for the total 100 trials.

LAW OF LARGE NUMBERS

A general principle according to which under certain very general conditions the simultaneous action of random factors leads to a result which is practically non-random. That the frequency of occurrence of a random event tends to become equal to its probability as the number of trials increases (a phenomenon which was probably first noted for games of chance) may serve as the first example of this principle.

At the turn of the 17th century J. Bernoulli demonstrated a theorem stating that, in a sequence of independent trials, in each of which the probability of occurrence of a certain A event has the same value $p, 0 < p < 1$, the relationship,

$$P\left\{\left|\frac{\mu_n}{n} - p\right| > \varepsilon\right\} \rightarrow 0$$

is valid for any $\varepsilon > 0$ if $n \rightarrow \infty$; here, μ_n is the number of occurrences of the event in the first *n* trials and *un*/*n* is the frequency of the occurrence. This Bernoulli theorem was extended by S. Poisson to the case of a sequence of independent trials in which the probability of the occurrence of an event *A*varies with the number of the trial. Let this probability in the k-th trial be $pk, k = 1, 2, \ldots$, and let,

$$\bar{P}n = \frac{p_1 + \ldots + p_n}{n}$$

Then, according to the Poisson theorem,

$$P\left\{\left|\frac{\mu_n}{n} - \bar{p}\right| > \varepsilon\right\} \rightarrow 0$$

for any $\varepsilon > 0$ if $n \rightarrow \infty$. A rigorous proof of this theorem was first given in 1846 by P.L. Chebyshev, whose method was quite different from that of Poisson and was based on certain extremal considerations, whereas Poisson deduced from an approximate formula for the above probability, based on a law of Gauss which at that time had not yet been proved rigorously. Poisson was the first to use the term "law of large numbers", by which he denoted his own generalization of the Bernoulli theorem.

A further natural extension of the Bernoulli and Poisson theorems is a consequence of the fact that the random variables may be represented as the sum,

$$\mu_n = X_1 + \ldots + X_n$$

of independent random variables, where $X_K = 1$ if A occurs in the k-th trial and $X_K = 0$ otherwise. The mathematical expectation $E(\mu_n/n)$ (which is identical with the average of the mathematical expectations $EX_K, 1 \leq k \leq n$) is P in the case of Bernoulli and is in $\bar{p}_n$ the case of Poisson. In other words, in both cases one deals with a deviation of the arithmetical average of the variables X_K from the arithmetical average of their mathematical expectations.

In his work "On average quantities", which appeared in 1867, Chebyshev established that the relationship,

$$p\left\{\left|\frac{X_1 + \ldots + X_n}{n} - \frac{EX_1 + \ldots + EX_n}{n}\right| > \varepsilon\right\} \to 0$$

is valid, for any if $\varepsilon > 0, n \to \infty$ for independent random variables $X_1, \ldots, X_n, \ldots,$ under fairly general assumptions. Chebyshev assumed that the mathematical expectations EX_K^2 are all bounded by the same constant, even though his proof makes it clear that the requirement of boundedness of the variances of $X_k, DX_k = EX_K^2 - (EX_K)^2$, or even the requirement,

$$B_n^2 = DX_1 + \ldots + DX_n = o(n^2), \qquad n \to \infty$$

is sufficient. It was thus shown by Chebyshev that the Bernoulli theorem is susceptible of extensive generalizations. A.A. Markov noted the possibility of further extensions and proposed to apply the term "law of large numbers" to all extensions of the Bernoulli theorem (and, in particular, to).

Chebyshev's method is based on a rigorous formulation of all the properties of the mathematical expectations and on the use of the so- called Chebyshev inequality in probability theory. For the probability it yields an estimate of the form

$$n^{-2}\varepsilon^{-2}\sum_{K=1}^{n} DX_k$$

This estimate may be replaced by a more exact one — but under more restrictive conditions, Subsequent proofs of the law of large numbers are all, to varying extents, developments of Chebyshev's method. Using an appropriate "truncation" of the random variables X_k(replacement of these variables by auxiliary variables,

$$X'_n k, \text{ viz. } X'_{n,k} = X_k \text{ if } X_k - EX_k| \le L_n$$

$$\text{and } X'_{n,K} = 0 \text{ if } X_k - EX_k| > L_n,$$

where L_n is a constant), Markov extended the law of large numbers to the case when the variance of the terms does not exist. He showed, for instance, that is valid if, for certain constants $\delta > 0$ and $L > 0$ for all *n*,

$$E|X_n - EX_n|^{1+\delta} < L$$

Khinchin's theorem (1929) is proved in a similar manner: If X_n have the same distribution and if EX_n exists, then the law of large numbers is valid.

It is possible to formulate more or less final versions of the law of large numbers for sums of independent random variables.

To do this, it is expedient to adopt a more general point of view, involving the concept of a sequence of asymptotically-constant random variables. The random variables of a sequence $Y_1,\ldots,Y_n,\ldots,$ are said to be asymptotically constant if there exists a sequence of constants $C_1,\ldots,C_n,\ldots,$ such that for any $\varepsilon > 0$, if $n > \infty$,

$$\mathrm{P}\{|Y_n - C_n| > \varepsilon\} \to 0$$

($i.e.\, Y_n - C_n$ converges to zero in probability). If is valid for C_n some, it is also valid for $C'_n = mYn$, where mY is the median of the random variable Y.

Furthermore, instead of the $X_1,\ldots,X_n,\ldots,$ sequence of independent random variables, one may also take a so-called triangular array:

$$X_{1,1},\dots,X_{1,k_1},$$
$$X_{2,1},\dots,X_{2,k_2},$$
$$\dots\dots,$$
$$X_{n,1},\dots,X_{n,k_n}:$$
$$\dots\dots$$

of random variables (the first subscript is the row number, while the second indicates the number of the variable within the row). The random variables of each individual row are assumed to be mutually independent. A sequence of partial sums is readily reduced to a triangular array if one assumes that,

$$k_1 = 1, k_2 = 2,\dots,\ X_{n,k} = X_k / n,$$

Let,

$$Y_n = X_{n,1+\dots+}X_{n,k_n}.$$

The general problem of the applicability of the law of large numbers to sums of independent random variables may then be stated as follows: Under what conditions are the sums Y_n asymptotically constant? This question was answered in 1928 by A.N. Kolmogorov. Assume, without loss of generality, that the medians of the variables $X_{n,\,k}$ are zero. Let,

$$X_{n,k} = X_{n,k}\ \text{if}\ |X_{n,}k| \le 1$$

and

$$X_{n,k} = 0\ \text{if}\ X_{n,\,k}| > 1.$$

The simultaneous fulfillment of the two conditions,

$$\sum_{k=1}^{kn} P\{|X_{n,k}| > 1\} \to 0 \quad if\ n \to \infty$$

and,

$$\sum_{k=1}^{k_n} E\bar{X}^2_{n,k} \to 0 \quad \text{if}\ n \to \infty,$$

is necessary and sufficient for the sums Y_n to be asymptotically constant. The sum,

$$\sum\nolimits_k^{k_n} 1EX_{n,k}$$

may be taken as C_n. That these conditions are sufficient is readily demonstrated by Chebyshev's method. If the mathematical expectations $EX_{n,k}$ exist, it is easy to find supplementary conditions under which it is permissible to choose $C_n = \mathrm{E}Y_n$, which yields necessary and sufficient conditions for the law of large numbers in the classical formulation. For a sequence of independent identically-distributed variables $\{X_n\}$ these conditions are reduced – in accordance with Khinchin's theorem quoted above – to the existence of the mathematical expectation. At the same time, the condition

$$n\mathrm{P}\{|X_1| > n\} \to 0 \text{ if } n \to \infty$$

is necessary and sufficient for the arithmetical averages Y_n to be asymptotically constant. Examples of cases in which condition is not met are easily found. Thus, the condition is not met if all X_n have a Cauchy distribution with density $1/\pi(1+x^2)$ (to which corresponds the characteristic function $\exp\{-|t|\}$). Here, the arithmetical averages,

$$Y_n = (X_1 + ... + X_n)/n$$

have the characteristic function $\exp\{-n|t/n|\} = \exp\{-|t|\}$ and, as a result, have the same distribution for any n as do the individual terms.

The most important cases to which the law of large numbers does not apply involve the times of return to the starting point in a random walk. Thus, in a symmetric Bernoulli random walk the time T_n passing prior to the n-th return to the starting point is the sum of the n independent random variables $X_1 .., X_n$ where X_1 is the time passed prior to the first return, is the time between the first and the second return, etc. The distribution of the variable $2T_{n/}\pi n^2$ converges as $n \to \infty$ to a non-degenerate limit distribution with density

$$P(x) = \frac{1}{\sqrt{2\pi}} e^{-1/2x} x^{-3/2} \textit{ for } x > 0$$

and zero for $x \leq 0$. Accordingly, in this case the distribution of the arithmetical average of X_i, *i.e.* $T_{n/}n$, is located, roughly speaking, on a segment of length of order n(whereas, in cases where the law of large numbers is applicable, it is located on segments of length) $_o(1)$. The applicability of the law of large numbers to sums of dependent variables (both in its classical and in its more general formulations) is connected, first and foremost, with the asymptotic independence of the random variables X_i and X_j as the difference between their subscripts $|i-j|$ increases. The corresponding theorems were first proved by Markov in 1907 for variables connected in a Markov chain. In fact, let be a homogeneous Markov chain with finite number of states, and let $X_1,...,X_n,...,$ all one-step transition probabilities be positive. Here, the asymptotic independence of X_j and X_i as $|i-j| \to \infty$ follows from the fact that the conditional distribution of X_j under a fixed value of X_i tends, as $n \to \infty$, to a limit which does not depend on the chosen value of X_i (Markov's ergodic theorem). The law of large numbers is deduced from this theorem: It is first established that, as $n \to \infty$,

$$\mathrm{E}\left(\frac{X_1+...+X_n}{n}-\alpha\right)^2 \to 0,$$

where $a = \lim_{k\to\infty} EX_k$; hence it follows that as $n \to \infty$,

$$\mathrm{P}\left\{\frac{X_1+...+X_n}{n}-\alpha\right| > \varepsilon\right\} \to 0,$$

The following, more general case, is governed by Bernstein's conditions: If $DX_j < L|R(X_i,X_j)| \le \phi(|i-j|)$,, where L is a constant, R is the correlation coefficient and $\phi(n)$ is a function which tends to zero as $n \to \infty$, then the law of large numbers is applicable to the variables $\{X_n\}$. For sequences $\{X_n\}$ which are stationary in the wide sense, the correlation condition may be weakened somewhat, by replacing it by

$$\lim_{n\to\infty} \frac{1}{n}\sum_{j=1}^{n} R_j = 0,$$

where,

$$R_j = R(X_i, X_{i+j})$$

The above-mentioned results may be extended in various ways. First, only convergence in probability was considered so far. Other types of convergence may also be considered, including convergence with probability one, mean-square convergence, etc. (in fact, many of the above conditions ensure mean-square convergence, which implies convergence in probability). The case of convergence with probability one is important, and for this reason has been especially named the strong law of large numbers.

Furthermore, many of these theorems can be applied, with suitable modifications, to random vectors having values in Euclidean spaces of any dimension, in a Hilbert space or in certain Banach spaces. For instance, if $\{X_n\}$ is a sequence of independent identically-distributed random vectors with values in a separable Banach space, and if $\mathrm{E}\|X_n\|$ (where $\|x\|$ is the norm of x) is finite, one has

$$\mathrm{P}\left\{\frac{X_1+\ldots+X_n}{n}-\mathrm{EX}_1\| > \varepsilon\right\} \to 0$$

for any $\varepsilon > 0$ if $n \to \infty$.

The law of large numbers, when considered in its most general form, is closely related to ergodic theorems (cf. Ergodic theorem). Clearly, many theorems are also applicable to the case of the average,

$$(1/T)\int_0^T X(t)dt,$$

where $X(t)$ is a random process depending on a continuous parameter. Finally, instead of the sums of random variables one may consider other symmetric functions of them. This was in fact done by A.Ya. Khinchin (1951–1955) to prove certain conclusions in statistical mechanics. The result obtained by him may be illustrated by the following example. Let $X_n,1, \ldots, X_{n,n}$ be the coordinates of a point uniformly distributed on the surface of the sphere

$$X_{n,1}^2+\ldots+X_{n,n}^2 = n\mu, \quad \mu > 0$$

The law of large numbers then applies to a wide class of symmetric functions $f(X_{n,1}, \ldots, X_{n,n})$ in the sense that as $n \to \infty$, their values are asymptotically constant (this is similar to the observation made in 1925 by P. Lévy to the effect that sufficiently regular functions of a very large number of variables are almost constant in a large part of their domain of definition).

BERNSTEIN INEQUALITY

Bernstein's inequality in probability theory is a more

precise formulation of the classical Chebyshev inequality in probability theory, proposed by S.N. Bernshtein in 1911; it permits one to estimate the probability of large deviations by a monotone decreasing exponential function. In fact, if the equations,

$$\mathrm{E}X_j = 0, \quad \mathrm{E}X_j^2 = B_j, \quad j = 1, \ldots, n,$$

hold for the independent random variables $X_1, \ldots, X_n$ with,

$$\mathrm{E}\left|X_j\right|^l \leq \frac{b_j}{2} H^{l-2} l!$$

(where $l > 2$ and H is a constant independent of), then the following inequality of Bernstein (where $r > 0$) is valid for the sum $S_n = X_1 + \ldots + X_n$:

$$\mathrm{P}\{|S_n| > r\} \leq 2\exp\left\{-\frac{r^2}{2(B_n + Hr)}\right\},$$

where $B_n = \sum b_j$. For identically-distributed bounded random variables $X_j(EX_j = 0,\ EX_j^2 = \sigma^2$, and $|X_j| \leq L$,) inequality takes its simplest form:

$$\mathrm{P}\{|S_n| > t\sigma\sqrt{n}\} \leq 2\exp\left\{-\frac{t^2}{2(1+\alpha/3)}\right\}$$

where $\alpha = Lt/\sqrt{n}\sigma$. A.N. Kolmogorov gave a lower estimate of the probability in. The Bernstein–Kolmogorov estimates are used, in particular, in proving the law of the iterated logarithm. Some idea of the accuracy of may be obtained by comparing it with the approximate value of the left-hand side of which is obtained by the central limit theorem in the form,

$$\frac{2}{\sqrt{2\pi}}\int_t^{\infty} e^{-u^2/2}\,du = \frac{2}{\sqrt{2\pi t}}\left(1 - \frac{\theta}{t^2}\right)e^{-t^2/2},$$

where $0 < \theta < 1$. Subsequent to 1967, Bernstein's inequalities were extended to include multi-dimensional and infinite-dimensional cases. Bernstein's inequality for the derivative of a trigonometric or algebraic polynomial gives an estimate of this derivative in terms of the polynomial itself. If $T_n(x)$ is a trigonometric polynomial of degree not n exceeding and if

$$M = \max_{0 \leq x \leq 2\pi} |T_n(x)|,$$

then the following inequalities are valid for all x:

$$|T_n^{(r)}(x)| \leq Mn^r, \quad r = 1, 2, \ldots.$$

Bernstein's inequality for an algebraic polynomial has the following form: If the polynomial,

$$P_n(x) = \sum_{k=0}^{n} \alpha_k x^k$$

satifies the condition,

$$|P_n(x)| \leq M, \quad \alpha \leq x \leq b$$

then its derivative $P_n'(x)$ has the property,

$$|P_n'(x)| \leq \frac{Mn}{\sqrt{(x-a)(b-x)}}, \quad \alpha < x < b,$$

which cannot be improved. As was noted by S.N. Bernshtein, this inequality is a consequence of the proof of the Markov inequality given by A.A. Markov.

Bernstein's inequalities are in fact employed in proving converse theorems in the theory of approximation of functions. There are a number of generalizations of Bernstein's inequality, in particular for entire functions in several variables.

12

Sampling Methods

The need for accurate information in order to make informed decisions has been emphasized throughout this online publication (refer to the sections on Data, information and statistics). Statistics are an important type of information and statistical agencies play an important part in the production of such information. In order to do this, the agencies rely on censuses, sample surveys and the use of administration records. These three sources, along with their advantages and disadvantages, have been explained.

Sampling allows statisticians to draw conclusions about a whole by examining a part. It enables us to estimate characteristics of a population by directly observing a portion of the entire population. Researchers are not interested in the sample itself, but in what can be learned from the survey—and how this information can be applied to the entire population.

It is essential that a sample survey be correctly defined and organized. If the wrong questions are posed to the wrong people, statisticians will not receive information that will be useful when applied to the entire population.

In the context of a national statistical agency like Statistics Canada, the following steps are needed to select a sample and ensure that this sample will fulfill its goals.

Establish the Survey's Objectives

The first step in planning a useful and efficient survey is to specify the objectives with as much detail as possible. Without objectives, the survey is unlikely to generate usable results. Clarifying the aims of the survey is critical to its ultimate success. The initial users and uses of the data should be identified at this stage. The pros and cons of a census versus a sample survey or the use of administrative records should be evaluated and a decision made as to the most appropriate method. (At this point, we will assume that a sample survey is the best way to proceed in order to obtain the information we need. This assumption will hold true for the remainder of the sample selection steps, even though many of the steps mentioned will also apply to the other methods.)

Define The Target Population

The target population is the total population for which the information is required. For example, if you were to conduct a survey about the most popular types of cars in Saskatchewan, then the target population would be every car in Saskatchewan. The units that make up the population must be described in terms of characteristics that clearly identify them.

Specifically, the target population is defined by the following characteristics:

- Nature of data required: about persons, hospitals, schools, etc.
- Geographic location: the geographic boundaries of the population have to be determined, as well as the level of geographic detail required for the survey estimate (by province, by city, etc.).
- Reference period: the time period covered by the survey.
- Other characteristics, such as socio-demographic characteristics (interest in a particular age group, for example) or type of industry.

Decide on the Data to be Collected

The data requirements of the survey must be established. To ensure that the requirements are operationally sound, the necessary data terms and definitions also need to be determined.

Set the Level of Precision

As mentioned in the section on Sampling error, there is a level of uncertainty associated with estimates coming from a sample. For example, if you are trying to estimate the average distance between home and school for students in your class of 25 from a sample of 5 persons, your estimate will depend on who the 5 sampled students are. If the 5 sampled students also live close to the school, the results will not be able to represent the class accurately. This sample-to-sample variation is what causes the sampling error. Statisticians can estimate the sampling error associated with a particular sampling plan, and try to minimize it. When designing a survey, the acceptable level of uncertainty in the survey estimates has to be established. This level depends on what the end use of the results will be and on the size of the overall budget. The bigger the budget, the more resources available, and thus, less chance for error. And if the end result is to serve a specific purpose, then the acceptable level of uncertainty would be smaller than an end result that is simply looking for general trends. The level of uncertainty will also be determined by the sample size. Increasing the sample size will decrease the sampling error. (If you sample 24 out of 25 students in your class, there will not be as much sample-to-sample variation as there would be if you only sampled 5 students from among the 25 possible samples.)

THE SAMPLE DESIGN

Once the objectives, guidelines and definitions have been worked out, the statistician can work on the survey plan. The survey plan is divided into three parts:

- Sample design: how the sample will be collected.
- Estimation techniques: how the results from the sample will be extended to the whole population.
- Measures of precision: how the sampling error will be measured.

The estimation techniques and measures of precision are discussed in a later section. For the moment, we will look at the sample design.

The following steps lead to the complete determination of the sample design:

1. Determine what the survey population will be (*e.g.*, students, men aged 20 to 35, newborn babies, etc.).
2. Choose the most appropriate survey time frame.
3. Define the survey units.
4. Establish the sample size (*e.g.*, a sample of 100 from a population of 1,000).
5. Select a sampling method.

The Survey Population

The target population must be defined early in the survey-designing process. This is the population for which information is required. However, some members of the population have to be excluded because of operational constraints: the high cost of collecting data in some remote areas, the difficulty of identifying and contacting certain components of the target population, etc.

For example, because it would be too difficult to locate and survey each car owned by every resident in Saskatchewan, a survey population of just the major cities and towns might be conducted instead. When some of the members of the target population are excluded, we call the included population the survey population or, what is sometimes called, an *observed population*. The target population is the population we want to observe while the survey population is the population we can observe.

The goal of this process is to have the survey population as close as possible to the target population. It is also very important that the users of the data be informed of the differences between the two populations, as the results of the survey will apply only to the survey population.

For example, a target population for a survey could be all Canadians aged 15 years and over (on a particular reference date), while the survey population could exclude residents of the Yukon, Nunavut and Northwest Territories, persons living on Aboriginal reserves, full-time members of the Canadian Armed Forces and residents of institutions.

These Canadians might be excluded for various reasons: to survey people in the territories might prove to be difficult and expensive, military personnel may not be available for surveying if they are out on a mission, etc. Using this example, about 2% of the target population would be excluded from the survey population.

The Survey Frame

The survey frame, also called the sampling frame, is the tool used to gain access to the population. There are two types of frames: list frames and area frames. A list frame is just a list of names and addresses that provide direct access to 'individuals' (*e.g.*, a list of hospitals, a list of restaurants, a list of students at a university). Area frames are a list of geographic areas that provide indirect access to individuals (*e.g.*, the neighbourhoods in a city). This type of access is called indirect because first, a list of geographic areas must be selected and then, access to individuals within each selected area must be worked out.

For instance, suppose that you were surveying a rural town in Quebec to see what percentage of residents are farmers. If you were provided with an area frame, then you would be able to locate which roads to visit, but you would still have to find out the names and addresses of the residents on each road. When there is no single frame that is appropriate, multiple frames can be used. Some sampling techniques using both types of frames will be discussed later. A good frame should be complete and up-to-date; no member of the survey population should be excluded from the frame or duplicated on the frame (represented more than once); and no unit that is not part of the population (*e.g.*, deceased persons) should be on the frame. The frame chosen will impact the selected survey population. For instance, if a list of telephone numbers is used to select a sample of households, then all households without telephones are excluded from the survey population.

The survey Units

There are three types of units that have to be accurately

identified in order to avoid problems during the selection, data collection and data analysis stages.

They are as follows:

- The sampling unit is part of the frame and therefore subject to being selected.
- The respondent unit or reporting unit provides the information needed by the survey.
- The unit of reference or unit of analysis—the unit about which information is provided—is used to analyse the survey results.

For example, in a survey about newborns in Edmonton, the sampling unit might be a household, the reporting unit one of the parents or a legal guardian, and the unit of reference the baby.

The sampling units may differ depending on the frame used. This is why the survey population, survey frame and survey units are defined in conjunction with one another.

The Sample Size

The level of precision needed for the survey estimates will impact the sample size. However, it is not as easy to determine the sample size as one may think. Generally, the actual sample size of a survey is a compromise between the level of precision to be achieved, the survey budget and any other operational constraints, such as budget and time. In order to achieve a certain level of precision, the sample size will depend, among other things, on the following factors:

- *The variability of the characteristics being observed*: If every person in a population had the same salary, then a sample of one person would be all you would need to estimate the average salary of the population. If the salaries are very different, then you would need a bigger sample in order to produce a reliable estimate.
- *The population size*: To a certain extent, the bigger the population, the bigger the sample needed. But once you reach a certain level, an increase in population no longer affects the sample size. For instance, the

necessary sample size to achieve a certain level of precision will be about the same for a population of one million as for a population twice that size.

- *The sampling and estimation methods*: Not all sampling and estimation methods have the same level of efficiency. You will need a bigger sample if your method is not the most efficient. But because of operational constraints and the unavailability of an adequate frame, you cannot always use the most efficient technique.

The Sampling Method

There are two types of sampling methods: probability sampling and non-probability sampling. The difference between them is that in probability sampling, every unit has a 'chance' of being selected, and that chance can be quantified. This is not true for non-probability sampling; every item in a population does not have an equal chance of being selected. The next section will describe features of both types of sampling and detail some of the methods related to each type.

PROBABILITY SAMPLING

Probability sampling involves the selection of a sample from a population, based on the principle of randomization or chance. Probability sampling is more complex, more time-consuming and usually more costly than non-probability sampling. However, because units from the population are randomly selected and each unit's probability of inclusion can be calculated, reliable estimates can be produced along with estimates of the sampling error, and inferences can be made about the population.

There are several different ways in which a probability sample can be selected. The method chosen depends on a number of factors, such as the available sampling frame, how spread out the population is, how costly it is to survey members of the population and how users will analyse the data. When choosing a probability sample design, your goal should be to minimize the sampling error of the estimates for the most important survey variables, while simultaneously minimizing the time and cost of conducting the survey.

Simple Random Sampling

In *simple random sampling,* each member of a population has an equal chance of being included in the sample. Also, each combination of members of the population has an equal chance of composing the sample. Those two properties are what defines simple random sampling. To select a simple random sample, you need to list all of the units in the survey population.

Example: To draw a simple random sample from a telephone book, each entry would need to be numbered sequentially. If there were 10,000 entries in the telephone book and if the sample size were 2,000, then 2,000 numbers between 1 and 10,000 would need to be randomly generated by a computer. Each number will have the same chance of being generated by the computer (in order to fill the simple random sampling requirement of an equal chance for every unit). The 2,000 telephone entries corresponding to the 2,000 computer-generated random numbers would make up the sample.

Simple random sampling can be done with or without replacement. A sample with replacement means that there is a possibility that the sampled telephone entry may be selected twice or more. Usually, the simple random sampling approach is conducted without replacement because it is more convenient and gives more precise results. For the purpose of these descriptions, when we discuss simple random sampling, we will refer to sampling without replacement.

Simple random sampling is the easiest method of sampling and it is the most commonly used. Advantages of this technique are that it does not require any additional information on the frame (such as geographic areas) other than the complete list of members of the survey population along with information for contact. Also, since simple random sampling is a simple method and the theory behind it is well established, standard formulas exist to determine the sample size, the estimates and so on, and these formulas are easy to use.

On the other hand, this technique makes no use of auxiliary information present on the frame (*i.e.*, number of employees in each business) that could make the design of the sample more efficient. And although it is easy to apply simple random sampling to small populations, it can be expensive and unfeasible for large populations because all elements must be identified and labeled prior to sampling. It can also be expensive if personal interviewers are required since the sample may be geographically spread out across the population.

A lottery draw is a good example of simple random sampling. For example, when a sample of 6 numbers is randomly generated from a population of 49, each number has an equal chance of being selected and each combination of 6 numbers has the same chance of being the winning combination. Even though people tend to avoid combinations such as 1-2-3-4-5-6, it has the same chance of being the winning set of numbers as the combination of 8-15-21-28-32-40.

Example: Suppose your school has 500 students and you need to conduct a short survey on the quality of the food served in the cafeteria. You decide that a sample of 10 students should be sufficient for your purposes. In order to get your sample, you assign a number from 1 to 500 to each student in your school. To select the sample, you use a table of randomly generated numbers. All you have to do is pick a starting point in the table (a row and column number) and look at the random numbers that appear there. In this case, since the data run into three digits, the random numbers would need to contain three digits as well. Ignore all random numbers after 500 because they do not correspond to any of the students in the school. Remember that the sample is without replacement, so if a number recurs, skip over it and use the next random number. The first 10 different numbers between 001 and 500 make up your sample.

Example: Imagine that you own a movie theatre and you are offering a special horror movie film festival next month. To decide which horror movies to show, you survey moviegoers asking them which of the listed movies are their favourites. To create the list of movies needed for your survey, you decide to sample 100 of the 1,000 best horror movies of all time. The horror movie population is divided evenly into classic movies (those filmed in or before 1969) and modern movies (those filmed in or later than 1970). One way of getting a sample would be to write out all of the movie titles on slips of paper and place them in an empty box. Then, draw out 100 titles and you will have your sample. By using this approach, you will have ensured that each movie had an equal chance of selection.

You can also calculate the probability of a given movie being selected. Since we know the sample size (n) and the total population (N), calculating the probability of being included in the sample becomes a simple matter of division:

Probability of being selected (same for each horror movie):

$= (n \div N) \times 100\%$

$= (100 \div 1,000) \times 100\%$

$= 10\%$

This means that every movie title on the list has a 10% or a 1 in 10 chance of being selected. You can see that that one disadvantage of simple random sampling (not the only disadvantage, but an important one) is that even if you know that the population is made up of 500 classic movies and 500 modern movies and you know each movie's release date from the sampling frame, no use is made of this information. This sample might contain 77 classic movies and 23 modern movies, which would not be representative of the whole horror movie population.

There are ways to overcome this problem (these will be briefly discussed in the Estimation section), but there are also ways to account for this information. (This will also be discussed later, under the section on Stratified sampling.)

Systematic sampling

Sometimes called *interval sampling, systematic sampling* means that there is a gap, or interval, between each selected unit in the sample.

In order to select a systematic sample, you need to follow these steps:

1. Number the units on your frame from 1 to N (where N is the total population size).
2. Determine the sampling interval (K) by dividing the number of units in the population by the desired sample size. For example, to select a sample of 100 from a population of 400, you would need a sampling interval of 400 ÷ 100 = 4. Therefore, K = 4. You will need to select one unit out of every four units to end up with a total of 100 units in your sample.
3. Select a number between one and K at random. This number is called *the random start* and would be the first number included in your sample. Using the sample above, you would select a number between 1 and 4 from a table of random numbers. If you choose 3, the third unit on your frame would be the first unit included in your sample; if you choose 2, your sample would start with the second unit on your frame.
4. Select every K^{th} (in this case, every fourth) unit after that first number. For example, the sample might consist of the following units to make up a sample of 100: 3 (the random start), 7, 11, 15, 19...395, 399 (up to N, which is 400 in this case).

Using the example above, you can see that with a systematic sample approach there are only four possible samples that can be selected, corresponding to the four possible random starts:

- 1, 5, 9, 13...393, 397
- 2, 6, 10, 14...394, 398
- 3, 7, 11, 15...395, 399
- 4, 8, 12, 16...396, 400

Each member of the population belongs to only one of the four samples and each sample has the same chance of being selected. From that, we can see that each unit has a one in four chance of being selected in the sample. This is the same probability as if a simple random sampling of 100 units was selected. The main difference is that with simple random sampling, any combination of 100 units would have a chance of making up the sample, while with systematic sampling, there are only four possible samples. From that, we can see how precise systematic sampling is compared with simple random sampling. The population's order on the frame will determine the possible samples for systematic sampling. If the population is randomly distributed on the frame, then systematic sampling should yield results that are similar to simple random sampling.

This method is often used in industry, where an item is selected for testing from a production line to ensure that machines and equipment are of a standard quality. For example, a tester in a manufacturing plant might perform a quality check on every 20th product in an assembly line. The tester might choose a random start between the numbers 1 and 20. This will determine the first product to be tested; every 20th product will be tested thereafter.

Interviewers can use this sampling technique when questioning people for a sample survey. The market researcher might select, for example, every 10th person who enters a particular store, after selecting the first person at random. The surveyor may interview the occupants of every fifth house on a street, after randomly selecting one of the first five houses.

Example: Imagine you have to conduct a survey on student housing for your university or college. Your school has an enrolment of 10,000 students and you want to take a systematic sample of 500 students. In order to do this, you must first determine what your sampling interval (K) would be:

Total population ÷ sample size = sampling interval:

$N \div n = K$

$= 10{,}000 \div 500$

$= 20$

To begin this systematic sample, all students would have to be assigned sequential numbers. The starting point would be chosen by selecting a random number between 1 and 20. If this number were 9, then the 9th student on the list would be selected along with every 20th student thereafter. The sample of students would be those corresponding to student numbers 9, 29, 49, 69...9,929, 9,949, 9,969 and 9,989.

In the examples used thus far, the sampling interval K was always a whole number, but this is not always the case. For example, if you want a sample of 30 from a population of 740, your sampling interval (or K) will be 24.7. In these cases, there are a few options to make the number easier to work with. You can round the number—either round it up to the nearest whole number or round it down. Rounding down will ensure that you select at least the number of units you originally wanted (and you can then delete some units to get the exact sample size you wanted). Techniques exist to adapt systematic sampling to the case where N (total population) is not a multiple of n (sample size), but still give a sample exactly the same as the n units. These techniques will not be discussed here.

The advantages of systematic sampling are that the sample selection cannot be easier (you only get one random number—the random start—and the rest of the sample automatically follows) and that the sample is distributed evenly over the listed population. The biggest drawback of the systematic sampling method is that if there is some cycle in the way the population is arranged on a list and if that cycle coincides in some way with the sampling interval, the possible samples may not be representative of the population. This can be seen in the following example:

Example: Suppose you run a large grocery store and have a list of the employees in each section. The grocery store is divided into the following 10 sections: deli counter, bakery, cashiers, stock, meat counter, produce, pharmacy, photo shop, flower shop and dry cleaning. Each section has 10 employees, including a manager (making 100 employees in total). Your list is ordered by section, with the manager listed first and then, the other employees by descending order of seniority.

If you wanted to survey your employees about their thoughts on their work environment, you might choose a small sample to answer your questions. If you use a systematic sampling approach and your sampling interval is 10, then you could end up selecting only managers or the newest employees in each section. This type of sample would not give you a complete or appropriate picture of your employees' thoughts.

Sampling with Probability Proportional to Size

Probability sampling requires that each member of the survey population have a chance of being included in the sample, but it does not require that this chance be the same for everyone. If there is information available on the frame about the size of each unit (*e.g.*, number of employees for each business) and if those units vary in size, this information can be used in the sampling selection in order to increase the efficiency. This is known as *sampling with probability proportional to size* (PPS). With this method, the bigger the size of the unit, the higher the chance it has of being included in the sample. For this method to bring increased efficiency, the measure of size needs to be accurate. This is a more complex sampling method that will not be discussed in further detail here.

Stratified Sampling

Using *stratified sampling,* the population is divided into homogeneous, mutually exclusive groups called strata, and then independent samples are selected from each stratum. Any of the sampling methods mentioned in this section (and others that exist) can be used to sample within each stratum. The sampling method can vary from one stratum to another. When simple random sampling is used to select the sample within each stratum, the sample design is called *stratified simple random sampling*. A population can be stratified by any variable that is available for all units on the sampling frame prior to sampling (*e.g.*, age, sex, province of residence, income, etc.).

Why do we need to create strata? There are many reasons, the main one being that it can make the sampling strategy more efficient. It was mentioned earlier that you need a larger sample to get a more accurate estimation of a characteristic that varies greatly from one unit to the other than for a characteristic that does not. For example, if every person in a population had the same salary, then a sample of one individual would be enough to get a precise estimate of the average salary.

This is the idea behind the efficiency gain obtained with stratification. If you create strata within which units share similar characteristics (*e.g.*, income) and are considerably different from units in other strata (*e.g.*, occupation, type of dwelling) then you would only need a small sample from each stratum to get a precise estimate of total income for that stratum. Then you could combine these estimates to get a precise estimate of total income for the whole population. If you were to use a simple random sampling approach in the whole population without stratification, the sample would need to be larger than the total of all stratum samples to get an estimate of total income with the same level of precision.

Stratified sampling ensures an adequate sample size for sub-groups in the population of interest. When a population is stratified, each stratum becomes an independent population and you will need to decide the sample size for each stratum.

Example: Suppose you want to estimate how many high school students have part-time jobs at the national level and also in each province. If you were to select a simple random sample of 25,000 people from a list of all high school students in Canada (assuming such a list was available for selection), you would end up on average with just a little over 100 people from Prince Edward Island, since they account for less than half of a per cent of the whole Canadian population. This sample would probably not be large enough for the kind of detailed analysis you had in mind. Stratifying your list by province, again assuming that this information is available, and then selecting a sample size for each province would allow you to decide on the exact sample size needed for that specific province. Thus, in order to get good representation of Prince Edward Island, you would use a larger sample than the one allotted to it by the simple random sampling approach.

Example: An Ontario school board wanted to assess student opinion on dropping Grade 13 from the secondary school programme. They decided to survey students from Elmsview High School. To ensure a representative sample of students from all grade levels, the school board used a stratified sampling technique.

In this case, the strata were the five grade levels (grades 9 to 13). The school board then selected a sample within each stratum. The students selected in this sample were extracted using simple random or systematic sampling, making up a total sample of 100 students.

Stratification is most useful when the stratifying variables are:

- Simple to work with,
- Easy to observe, and
- Closely related to the topic of the survey.

Cluster Sampling

Sometimes it is too expensive to spread a sample across the population as a whole. Travel costs can become expensive if interviewers have to survey people from one end of the country to the other. To reduce costs, statisticians may choose a *cluster sampling* technique.

Cluster sampling divides the population into groups or clusters. A number of clusters are selected randomly to represent the total population, and then all units within selected clusters are included in the sample. No units from non-selected clusters are included in the sample—they are represented by those from selected clusters. This differs from stratified sampling, where some units are selected from each group. Examples of clusters are factories, schools and geographic areas such as electoral subdivisions. The selected clusters are used to represent the population.

Example: Suppose you are a representative from an athletic organization wishing to find out which sports Grade 11 students are participating in across Canada. It would be too costly and lengthy to survey every Canadian in Grade 11, or even a couple of students from every Grade 11 class in Canada. Instead, 100 schools are randomly selected from all over Canada. These schools provide clusters of samples. Then every Grade 11 student in all 100 clusters is surveyed. In effect, the students in these clusters represent all Grade 11 students in Canada.

Example: Imagine that the municipal council of a small city wants to investigate the use of health care services by residents.

First, the council requests from Statistics Canada electoral subdivision maps that identify and label each city block. From these maps, the council creates a list of all city blocks. This list will serve as the sampling frame. Every household in that city belongs to a city block, and each city block represents a cluster of households. The council randomly picks a number of city blocks. Using the simple random sample approach, then the council creates a list of all households in the selected city blocks; these households make up the survey sample.

As mentioned, cost reduction is a reason for using cluster sampling. It creates 'pockets' of sampled units instead of spreading the sample over the whole territory. Another reason is that sometimes a list of all units in the population (a requirement when conducting simple random sample, systematic sample or sampling with probability proportional to size) is not available, while a list of all clusters is either available or easy to create.

In most cases, the main drawback is a loss of efficiency when compared with simple random sampling. It is usually better to survey a large number of small clusters instead of a small number of large clusters. This is because neighbouring units tend to be more alike, resulting in a sample that does not represent the whole spectrum of opinions or situations present in the overall population. In the two previous examples, students in the same school tend to participate in the same types of sports (depending on the facilities available at their school); similarly, elderly people have a tendency to live in specific neighbourhoods and to be heavy users of health services.

Another drawback to cluster sampling is that you do not have total control over the final sample size. Since not all schools have the same number of Grade 11 students and city blocks do not all have the same number of households, and you must interview every student or household in your sample, the final size may be larger or smaller than you expected.

Multi-stage Sampling

Multi-stage sampling is like the cluster method, except that it involves picking a sample from within each chosen cluster, rather than including all units in the cluster. This type of sampling requires at least two stages. In the first stage, large groups or clusters are identified and selected. These clusters contain more population units than are needed for the final sample. In the second stage, population units are picked from within the selected clusters (using any of the possible probability sampling methods) for a final sample. If more than two stages are used, the process of choosing population units within clusters continues until there is a final sample.

Example: A cluster sample would choose 100 schools and then interview every Grade 11 student from those schools. Instead in multi-stage sampling, you could select more schools, get a list of all Grade 11 students from these selected schools and select a random sample (*e.g.*, simple random sample) of students from each school. This would be a two-stage sampling design.

You could also get a list of all Grade 11 classes in the selected schools, pick a random sample of classes from each of those schools, get a list of all the students in the selected classes and finally select a random sample of students from each class. This would be a three-stage sampling design. Each time we add a stage, the process becomes more complex. Now imagine that each school has on average 80 Grade 11 students. Cluster sampling would then give your organization a sample of about 8,000 students (100 schools x 80 Grade 11 students). If you wanted a bigger sample, you could select schools with more students; and for a smaller sample you could select schools with fewer students.

One way to control the sample size would be to stratify the schools into large, medium and small sizes (in terms of the number of Grade 11 students) and select a sample of schools from each stratum. This is called *stratified cluster sampling*.

With a three-stage design, you could select a sample of 400 schools, then select two Grade 11 classes per school (assuming that there are two or more Grade 11 classes per school). Finally, you could select 10 students per class. This way, you still end up with a sample of about 8,000 students (400 schools x 2 classes x 10 students), but the sample will be more spread out.

You can see from this example that with multi-stage sampling, you still have the benefit of a more concentrated sample for cost reduction. However, the sample is not as concentrated as other clusters and the sample size is still bigger than for a simple random sample size. Also, you do not need to have a list of all of the students in the population. All you need is a list of the classes from the 400 schools and a list of the students from the 800 classes. Admittedly, more information is needed in this type of sample than what is required in cluster sampling. However, multi-stage sampling still saves a great amount of time and effort by not having to create a list of all of the units in a population.

Multi-phase Sampling

A *multi-phase sample* collects basic information from a large sample of units and then, for a subsample of these units, collects more detailed information. The most common form of multi-phase sampling is two-phase sampling (or double sampling), but three or more phases are also possible.

Multi-phase sampling is quite different from multi-stage sampling, despite the similarities in name. Although multi-phase sampling also involves taking two or more samples, all samples are drawn from the same frame and at each phase the units are structurally the same. However, as with multi-stage sampling, the more phases used, the more complex the sample design and estimation will become.

Multi-phase sampling is useful when the frame lacks auxiliary information that could be used to stratify the population or to screen out part of the population.

Example: Suppose that an organization needs information about cattle farmers in Alberta, but the survey frame lists all types of farms—cattle, dairy, grain, hog, poultry and produce. To complicate matters, the survey frame does not provide any auxiliary information for the farms listed there. A simple survey could be conducted whose only question is "Is part or all of your farm devoted to cattle farming?" With only one question, this survey should have a low cost per interview (especially if done by telephone) and, consequently, the organization should be able to draw a large sample. Once the first sample has been drawn, a second, smaller sample can be extracted from among the cattle farmers and more detailed questions asked of these farmers. Using this method, the organization avoids the expense of surveying units that are not in this specific scope (*i.e.*, non-cattle farmers).

Multi-phase sampling can be used when there is insufficient budget to collect information from the whole sample, or when doing so would create excessive burden on the respondent, or even when there are very different costs of collection for different questions on a survey.

Example: A health survey asks participants some basic questions about their diet, smoking habits, exercise routines and alcohol consumption. In addition, the survey requires that respondents subject themselves to some direct physical tests, such as running on a treadmill or having their blood pressure and cholesterol levels measured.

Filling out questionnaires or interviewing participants are relatively inexpensive procedures, but the medical tests require the supervision and assistance of a trained health practitioner, as well as the use of an equipped laboratory, both of which can be quite costly. The best way to conduct this survey would be to use a two-phase sample approach. In the first phase, the interviews are performed on an appropriately sized sample. From this sample, a smaller sample is drawn. This second sample will take part in the medical tests.

NON-PROBABILITY SAMPLING

The difference between *probability* and *non-probability sampling* has to do with a basic assumption about the nature of the population under study. In probability sampling, every item has a chance of being selected. In non-probability sampling, there is an assumption that there is an even distribution of characteristics within the population. This is what makes the researcher believe that any sample would be representative and because of that, results will be accurate. For probability sampling, randomization is a feature of the selection process, rather than an assumption about the structure of the population.

In non-probability sampling, since elements are chosen arbitrarily, there is no way to estimate the probability of any one element being included in the sample. Also, no assurance is given that each item has a chance of being included, making it impossible either to estimate sampling variability or to identify possible bias. Reliability cannot be measured in non-probability sampling; the only way to address data quality is to compare some of the survey results with available information about the population. Still, there is no assurance that the estimates will meet an acceptable level of error. Statisticians are reluctant to use these methods because there is no way to measure the precision of the resulting sample.

Despite these drawbacks, non-probability sampling methods can be useful when descriptive comments about the sample itself are desired. Secondly, they are quick, inexpensive and convenient. There are also other circumstances, such as in applied social research, when it is unfeasible or impractical to conduct probability sampling. Statistics Canada uses probability sampling for almost all of its surveys, but uses non-probability sampling for questionnaire testing and some preliminary studies during the development stage of a survey.

Most non-sampling methods require some effort and organization to complete, but others, like convenience sampling, are done casually and do not need a formal plan of action.

Convenience or Haphazard Sampling

Convenience sampling is sometimes referred to as *haphazard* or *accidental sampling*. It is not normally representative of the target population because sample units are only selected if they can be accessed easily and conveniently. There are times when the average person uses convenience sampling. A food critic, for example, may try several appetizers or entrees to judge the quality and variety of a menu. And television reporters often seek so-called 'people-on-the-street interviews' to find out how people view an issue. In both these examples, the sample is chosen randomly, without use of a specific survey method.

The obvious advantage is that the method is easy to use, but that advantage is greatly offset by the presence of bias. Although useful applications of the technique are limited, it can deliver accurate results when the population is homogeneous. For example, a scientist could use this method to determine whether a lake is polluted. Assuming that the lake water is well-mixed, any sample would yield similar information. A scientist could safely draw water anywhere on the lake without fretting about whether or not the sample is representative.

Examples of convenience sampling include:

- The female moviegoers sitting in the first row of a movie theatre
- The first 100 customers to enter a department store
- The first three callers in a radio contest.

Volunteer Sampling

As the term implies, this type of sampling occurs when people volunteer their services for the study. In psychological experiments or pharmaceutical trials (drug testing), for example, it would be difficult and unethical to enlist random participants from the general public. In these instances, the sample is taken from a group of volunteers. Sometimes, the researcher offers payment to entice respondents. In exchange, the volunteers accept the possibility of a lengthy, demanding or sometimes unpleasant process.

Sampling voluntary participants as opposed to the general population may introduce strong biases. Often in opinion polling, only the people who care strongly enough about the subject one way or another tend to respond. The silent majority does not typically respond, resulting in large selection bias. Television and radio media often use call-in polls to informally query an audience on their views. The Much Music television channel uses this kind of survey in their CombatZone programme. The programme asks viewers to cast a vote for one of two music videos by telephone, e-mail or through their online website.

Oftentimes, there is no limit imposed on the frequency or number of calls one respondent can make. So, unfortunately, a person might be able to vote repeatedly. It should also be noted that the people who contribute to these surveys might have different views than those who do not.

Judgement Sampling

This approach is used when a sample is taken based on certain judgements about the overall population. The underlying assumption is that the investigator will select units that are characteristic of the population. The critical issue here is objectivity: how much can judgment be relied upon to arrive at a typical sample? *Judgement sampling* is subject to the researcher's biases and is perhaps even more biased than haphazard sampling. Since any preconceptions the researcher may have are reflected in the sample, large biases can be introduced if these preconceptions are inaccurate.

Statisticians often use this method in exploratory studies like pre-testing of questionnaires and focus groups. They also prefer to use this method in laboratory settings where the choice of experimental subjects (*i.e.*, animal, human, vegetable) reflects the investigator's pre-existing beliefs about the population. One advantage of judgement sampling is the reduced cost and time involved in acquiring the sample.

Quota Sampling

This is one of the most common forms of non-probability sampling. Sampling is done until a specific number of units (quotas) for various sub-populations have been selected. Since there are no rules as to how these quotas are to be filled, *quota sampling* is really a means for satisfying sample size objectives for certain sub-populations. The quotas may be based on population proportions. For example, if there are 100 men and

100 women in a population and a sample of 20 are to be drawn to participate in a cola taste challenge, you may want to divide the sample evenly between the sexes—10 men and 10 women. Quota sampling can be considered preferable to other forms of non-probability sampling (*e.g.*, judgement sampling) because it forces the inclusion of members of different sub-populations.

Quota sampling is somewhat similar to stratified sampling in that similar units are grouped together. However, it differs in how the units are selected. In probability sampling, the units are selected randomly while in quota sampling it is usually left up to the interviewer to decide who is sampled. This results in selection bias. Thus, quota sampling is often used by market researchers (particularly for telephone surveys) instead of stratified sampling, because it is relatively inexpensive and easy to administer and has the desirable property of satisfying population proportions. However, it disguises potentially significant bias.

As with all other non-probability sampling methods, in order to make inferences about the population, it is necessary to assume that persons selected are similar to those not selected. Such strong assumptions are rarely valid.

Example: The student council at Cedar Valley Public School wants to gauge student opinion on the quality of their extracurricular activities. They decide to survey 100 of 1,000 students using the grade levels (7 to 12) as the sub-population. The table below gives the number of students in each grade level.

Table. Number of Students Enrolled at Cedar Valley Public School, by Grade

Grade level of students sample	Number of students (%)	Percentage	Quota of studentsin of 100
7	150	15	15
8	220	22	22
9	160	16	16
10	150	15	15

11	200	20	20
12	120	12	12
Total	1,000	100	100

The student council wants to make sure that the percentage of students in each grade level is reflected in the sample. The formula is:

Percentage of students in Grade 10
= (number of students ÷ number of students) x 100%
$= (150 \div 1{,}000) \times 100$
= 15%

Since 15% of the school population is in Grade 10, 15% of the sample should contain Grade 10 students. Therefore, use the following formula to calculate the number of Grade 10 students that should be included in the sample:

Sample of Grade 10 students
$= (15\% \text{ of } 100) \times 100$
$= 0.15 \times 100$
= 15 students

The main difference between stratified sampling and quota sampling is that stratified sampling would select the students using a probability sampling method such as simple random sampling or systematic sampling. In quota sampling, no such technique is used. The 15 students might be selected by choosing the first 15 Grade 10 students to enter school on a certain day, or by choosing 15 students from the first two rows of a particular classroom. Keep in mind that those students who arrive late or sit at the back of the class may hold different opinions from those who arrived earlier or sat in the front.

The main argument against quota sampling is that it does not meet the basic requirement of randomness. Some units may have no chance of selection or the chance of selection may be unknown. Therefore, the sample may be biased.

It is common, but not necessary, for quota samples to use random selection procedures at the beginning stages, much in the same way as probability sampling does. For instance, the first step in multi-stage sampling would be randomly selecting the geographic areas. The difference is in the selection of the units in the final stages of the process.

In multi-stage sampling, units are based on up-to-date lists for selected areas and a sample is selected according to a random process. In quota sampling, by contrast, each interviewer is instructed on how many of the respondents should be men and how many should be women, as well as how many people should represent the various age groups. The quotas are therefore calculated from available data for the population, so that the sexes, age groups or other demographic variables are represented in the correct proportions.

But within each quota, interviewers may fail to secure a representative sample of respondents. For example, suppose that an organization wants to find out information about the occupations of men aged 20 to 25. An interviewer goes to a university campus and selects the first 50 men aged 20 to 25 that she comes across and who agree to participate in her organization's survey. However, this sample does not mean that these 50 men are representative of all men aged 20 to 25.

Quota sampling is generally less expensive than random sampling. It is also easy to administer, especially considering the tasks of listing the whole population, randomly selecting the sample and following-up on non-respondents can be omitted from the procedure. Quota sampling is an effective sampling method when information is urgently required and can be carried out independent of existing sampling frames. In many cases where the population has no suitable frame, quota sampling may be the only appropriate sampling method.

ESTIMATION

As we now know, the goal of conducting surveys is to obtain information about a particular population. When the sample has been selected and the information collected and processed, there still remains the task of linking the information gathered from the sample back to the overall population.

Estimation is the process of determining a likely value for a variable in the survey population, based on information collected from the sample. Researchers are usually interested in looking at estimates of many statistics—totals, averages and proportions being the most frequent—for different variables. For example, a sample survey could be used to produce any of the following statistics: estimates for the proportion of smokers among all people aged 15 to 24 in the population; the average earnings of men and women with a university degree; or the total number of cars possessed by the whole survey population.

Underpinning the estimation process is the sampling weight of a unit, which indicates the number of units in the population (including the sampling weight) that are represented by this sampled unit. The sampling weight is the inverse of the unit's probability of selection.

Example: Suppose that the city of Winnipeg has decided to award bus travellers with free one-year bus passes as a way of promoting its services. A simple random sample of 10 people is selected from the 30 passengers on a city bus. Since simple random sampling gives equal probability of selection to every member of the population (in this case, all passengers on the bus), each passenger had one chance out of three of being selected. This translates into a sampling weight of three for every selected unit. This means that each person in the sample represents three persons in the population—himself or herself, plus two other persons

To estimate this sampling weight, one could take the survey information for the 10 selected passengers and copy it three times to create an artificial population of 30. Totals, averages or proportions for the real population could then be estimated by the corresponding statistics computed using the artificial population. Instead of doing this, however, survey statisticians attach a sampling weight to each unit in the sample and take this weight into account when estimating.

If one person in a sample (with a sampling weight of 18) had blue eyes and brown hair, then it is as if a total of 18 people in the population had blue eyes and brown hair.

Example: You are conducting a survey to determine the total number of people living on your street and the average number of cars owned by each household. You decide to select a systematic sample of 5 households from the 20 on your street and intend to use that sample to estimate the totals you are looking for. The following table summarizes the information that you gathered during your interviews with the sampled households:

Table. Sample of households on Redwood Street

Household number	Number of persons	Number of cars	Probability of selection	Sampling weight
1	1	0	1/4	4
2	4	2	1/4	4
3	2	1	1/4	4
4	2	1	1/4	4
5	3	2	1/4	4

The selection probability of 1 in 4 comes from the fact that systematic sampling gives an equal chance of being selected to each household on your street. The sampling weight of 4 is just the inverse of that probability. When estimating, you have to look at the characteristics of each sampled household. In this case, it is decided that 4 households from the population of 20 on your street have the same characteristics.

In order to estimate the total number of persons living on your street, you have to multiply the number of persons in a household by the number of households in that sampling weight, then add up all the final numbers. For example, there are 4 one-person households (represented by Household number 1), 4 four-person households, 8 two-person households (four households represented by Household number 3 and four households represented by Household number 4) and 4 three-person households. The estimation of the total number of persons would then be:

Estimated number of persons living on your street:

$= (4\times 1) + (4\times 4) + (8\times 2) + (4\times 3)$

= 48 people

To estimate the average number of cars per household, you proceed in the same manner. Get an estimate of the total number of cars owned by households on your street and then, divide the estimate by the actual number of households on the street. For example, there are 4 households without a car (represented by Household number 1), 8 households with two cars (represented by Household number 2 and Household number 5), 8 households with one car each (represented by Household number 3 and Household number 4).

Estimated number of cars:

$= (4\times 0) + (8\times 2) + (8\times 1)$

$= 24$ cars

Estimated average:

$= 24 \div 20$

$= 1.2$ cars per household

Self-weighting Designs

It is not always the case that all sampled units had the same sampling weight. Some designs give unequal probability of selection to units, resulting in units within the same sample having different sampling weights. Answers from one household or business could represent the answers for 200 units of the population, while the answers from another could represent only 50 units in the population.

When every unit in the sample has the same sampling weight, the sampling design is said to be self-weighted. This kind of design is time-saving and operationally convenient, particularly for large samples. Because every unit has the same weight, those weights can be ignored when estimating averages and proportions. The average for the sample gives an appropriate estimate of the average for the whole population.

Simple random sampling and systematic sampling are examples of self-weighted designs. In that sense, calculations could have been made easier in Example 2. For instance, to estimate the average number of cars per household in the population, we could have used the same average as the one used ir. the sample. The 5 sampled households own a total of

6 cars, an average of 1.2 cars per household. This is the same result as that obtained using the sampling weight procedure.

Adjusting The Weights

Sometimes, the sampling weights are adjusted prior to estimation.

There are basically two reasons for weight adjustment:

- Adjusting for non-response: Using sampling weights for estimation works fine when you have been able to interview all selected units. In Example 2, if two of the five sampled households refused to answer or were unavailable at the time of the survey, you would only have answers for three households, thus representing only 12 of the 20 households on the street. The two non-responding units represent four households each. This means that we have no information on the number of persons or cars for 8 households on your street. In order to adjust for that, survey statisticians usually increase the weights of responding units to account for the loss of representativeness caused by non-response. The goal would be to use only the 3 units for which we have information, but still represent the 20 households on the street.
- Adjusting for external information: Sometimes, we know the actual total for one or more variables measured in the sample. In Example of the Probability sampling section, a population of the 1,000 best horror movies was equally divided into 500 classic movies and 500 modern movies. Even though you knew this prior to sampling, you decided to select a simple random sample of 100 movies and ended up with 77 classic movies and 23 modern movies. Each of these movies has a weight of 10 (because you selected 1 movie out of every 10 titles). Using the answers from the survey and the sampling weight, your sample would represent a population of 770 classic movies and 230 modern movies. This

could lead to inaccurate estimates. One solution would be to decrease the weight of every sampled classic and increase the weight of every sampled modern movie so that your sample gives an estimate of 500 classics and 500 modern films in the population. This should reduce the distortion caused by a 'bad' sample.

Of course, stratifying by release date prior to sampling would have solved this problem. However, in a lot of cases, we have totals at the population level, but we don't know the attribute of each unit on the sampling frame. For example, from the Census of Population, we know how many men and women there are in a specific city, but all we have for sampling is a list of households. Thus, stratifying our population by sex would not be possible. Demographic projections by age and sex for each province are often used in social surveys to adjust sampling weights. The weights adjusted for non-response and/or external counts are used for estimation, in the same way as the sampling weight was used in Example.

Other Estimation Methods

Using the weights to inflate the sample results is not the only estimation method that exists, but it is the simplest one and the only one that we will cover. Nevertheless, it is important to know that there exist some other methods that could lead to more precise estimates (*e.g.*, using auxiliary information). The estimation process has to take into account the sampling design that was used. Otherwise, the resulting estimates could be severely biased.

Estimating The Sampling Error

As mentioned before, any estimates derived from samples are subject to what is called the sampling error. This comes from the fact that only a part of the population was observed, instead of the whole. A different sample could have come up with different results. The amount of variation that exists among the estimates from the different possible samples is what makes the sampling error. (There are roughly 14 million different combinations of 6 numbers from 1 to 49, so imagine how many ways there are to select a sample of 25,000 Canadian households!) Of course, this sampling error is unknown, since we would need to know the answer for each unit of the population in order to calculate it. Nevertheless, it can be estimated by using the survey data. The extent of the sampling error depends on many things, including the sampling method, the estimation method, the sample size and the variability of the estimated characteristic. This is why each sample estimate has its own sampling error. This error should thus be approximated for each estimated total, average, proportion, etc. produced by the survey.

Examples of Estimation Using an Simple Random Sampling Design

Simple random sampling is the simplest of all sampling methods. Estimation using the simple random sampling method has been studied extensively. There are simple formulas to estimate the sampling error for many statistics when simple random sampling is used, especially since it is a self-weighting design. We present here the most common estimator for a population average (mean) and total, under simple random sampling.

Estimation of The Population Mean

In a simple random sample, the estimate of the population mean is identical to the mean of the sample:

$$\hat{x} = \frac{\sum x}{n}$$

where,

x = an observed value $\hat{x}$

= estimate of the population mean

$\sum$ x = sum of all observed x values in the sample

n = number of observations in the sample.

Note: Lowercase x and n should be used if you are referring to a sample survey and upper case X and N should be used when referring to a population.

If the sample results have been summarized in a frequency

table, then the estimate for the population mean is the same as the sample.

Thus,

$$\hat{x} = \frac{\sum xf}{\sum f}$$

where,

x = An observed value

f = The frequency of the value (the number of times that this value have been observed in the sample) $\hat{x}$

= Estimate of the population mean $\sum xf$

= Sum of all observed xf values (the product of the observed values times its frequency) in the sample $\sum f$

= Sum of the frequencies in the sample.

Example: A farmer randomly selects 10 eggs from a gross of 12 dozen eggs (144 eggs) he finds in his hen house. He carefully weighs each egg.

The following weights were recorded in grams:

0.75, 0.70, 0.55, 0.50, 0.60, 0.65, 0.75, 0.65, 0.75, 0.50

What is the mean weight of the gross of eggs?

Using the above formula, we can determine the mean weight of all of the eggs:

$$\hat{x} = \frac{\sum x}{n}$$

$$= \frac{6.4}{10}$$

= 0.64 grammes

Estimation of The Population Total

For a simple random sample, the estimation formula of a total for the population is:

$$\hat{x} = N\frac{\sum x}{n}$$

where,

x = An observed value $\hat{x}$

= Estimated population total $\sum x$

= Sum of all observed x values in the sample

n = Number of observations in the sample

N = Total number of observations in the population.

It is just the estimate for the mean value multiplied by the number of units in the population. In the previous example, the mean weight of an egg is 0.64 grams, so it is logical to think that the total weight of the 144 eggs would be 92.16 grams (144 x 0.64 = 92.16 grams).

13

Measure and Probability Theory

The evolution of probabil ity theory was based more on intuition rather than mathematical axioms during its early development. In 1933, A. N. Kolmogorov provided an axiomatic basis for probability theory and it is now the universally accepted model. There are certain 'non commutative' versions that have their origins in quantum mechanics, see for instance K. R. Parthasarathy, that are generalizations of the Kolmogorov Model. We shall however use exclusively Kolmogorov's framework.

The basic intuition in probability theory is the notion of randomness. There are experiments whose results are not predicatable and can be determined only after performing it and then observing the outcome. The simplest familiar examples are, the tossing of a fair coin, or the throwing of a balanced die. In the first experiment the result could be either a head or a tail and the throwing of a die could result in a score of any integer from 1 through 6. These are experiments with only a finite number of alternate outcomes. It is not difficult to imagine experiments that have countably or even uncountably many alternatives as possible outcomes.

Abstractly then, there is a space Ω of all possible outcomes and each individual outcome is represented as a point ω in that space Ω. Subsets of Ω are called events and each of them corresponds to a collection of outcomes. If the outcome ω is in the subset A, then the event A is said to have occurred. For example in the case of a die the set $A = \{1, 3, 5\}$ M Ω corresponds to the event 'an odd number shows up'. With this terminology it is clear that union of sets corresponds to 'or', intersection to 'and', and complementation to 'negation'. One would expect that probabilities should be associated with each outcome and there should be a 'Probability Function' $f(\omega)$ which is the probabilty that ω occurs. In the case of coin tossing we may expect $\Omega = \{H, T\}$ and,

$$f(T) = f(H) = \frac{1}{2}.$$

Or in the case of a die,

$$f(1) = f(2) = ... = f(6) = \frac{1}{6}.$$

Since 'Probability' is normalized so that certainty corresponds to a Probability of 1, one expects,

$$\sum_{\omega \in \Omega} f(\omega) = 1.$$

If Ω is uncountable this is a mess. There is no reasonable way of adding up an uncountable set of numbers each of which is 0. This suggests that it may not be possible to start with probabilities associated with individual outcomes and build a meaningful theory. The next best thing is to start with the notion that probabilities are already defined for events. In such a case, $P(A)$ is defined for a class B of subsets $A \subset \Omega$. The question that arises naturally is what should B be and what properties should $P(\cdot)$ defined on B have? It is natural to demand that the class B of sets for which probabilities are to be defined satisfy the following properties:

The whole space Ω and the empty set Φ are in B. For any two sets A and B in B, the sets $A \cup B$ and $A \cap B$ are again in B. If $A \in B$, then the complement A^c is again in B. Any class of sets satisfying these properties is called a *field*.

Definition: A 'probability' or more precisely 'a finitely additive probability measure' is a nonnegative set function $P(\cdot)$ defined for sets $A \in B$ that satisfies the following properties:

$$P(A) \geq 0 \text{ for all } A \in B,$$

$$P(\Omega) = 1 \text{ and } P(\Phi) = 0.$$

If $A \in B$ and $B \in B$ are disjoint then,

$$P(A \cup B) = P(A) + P(B).$$

In particular,

$$P(A^c) = 1 - P(A)$$

for all $A \in B$.

A condition which is some what more technical, but important from a mathematical viewpoint is that of countable additivity. The class B, in addition to being a field is assumed to be closed under countable union (or equivalently, countable intersection); *i.e.* if $A_n \in B$ for every n, then $A = \bigcup_n A_n \in B$. Such a class is called a σ-field. The 'probability' itself is presumed to be defined on a σ-field B.

Definition: A set function P defined on a σ-field is called a 'countably additive probability measure' if in addition to satsfying equations, it satisfies the following countable additivity property: for any sequence of pairwise disjoint sets A_n with $A = \bigcup_n A_n$

$$P(A) = \sum_n P(A_n).$$

A probability measure P on a field F is said to be countably additive on F, if for any sequence $A_n \in F$ with $A_n \downarrow \Phi$, we have $P(A_n) \downarrow 0$. The σ-field in the above exercise is called the σ-field generated by F.

CONSTRUCTION OF MEASURES

The following theorem is important for the construction of countably additive probability measures. A detailed proof of this theorem, as well as other results on measure and integration, can be found or in any one of the many texts on real variables. In an effort to be complete we will sketch the standard proof.

Caratheodory Extension Theorem

Theorem: Any countably additive probabilty measure P on a field F extends uniquely as a countably additive probability measure to the σ-field B generated by F.

Proof: The proof proceeds along the following steps:

Step 1: Define an object P^* called the *outer measure* for all sets A.

$$P^*(A) = \inf_{\cup_j A_j \supset A} \sum_j P(A_j)$$

where the infimum is taken over all countable collections $\{A_j\}$ of sets from F that cover A. Without loss of generality we can assume that $\{A_j\}$ are disjoint. (Replace A_j by,

$$\left(\bigcap_{i=1}^{j-1} A_i^c\right) \cap A_j).$$

Step 2: Show that P has the following properties:*

1. P^* is countably sub-additive, *i.e.*

$$P^*\left(\bigcup_j A_j\right) \leq \sum_j P^* A_j$$

2. For A Î F, $P^*(A) \leq P(A)$. (Trivial)
3. For A Î F, $P^*(A) \geq P(A)$. (Need to use the countable additivity of P on F)

Step 3: Define a set E to be measurable if:

$$P^*(A) \geq P^*(A \cap E) + P^*(A \cap E^c)$$

holds for all sets A, and establish the following properties for the class M of measurable sets. The class of measurable sets M is a σ-field and P^* is a countably additive measure on it.

Step 4: Finally show that $M \supset F$. This implies that $M \supset B$ and P^* is an extension of P from F to B. Uniqueness is quite simple. Let P_1 and P_2 be two countably additive probability measures on a σ-field B that agree on a field $F \subset B$. Let us define $A = \{A: P_1(A) = P_2(A)\}$. Then A is a monotone class *i.e.*, if $A_n \in A$ is increasing (decreasing) then,

$$\bigcup_n A_n \left(\bigcap_n A_n\right) \in A.$$

Uniqueness will follow from the following fact left as an excercise.

Exercise: The smallest monotone class generated by a field is the same as the σ-field generated by the field. It now follows that A must contain the σ-field generated by F and that proves uniqueness.

The extension Theorem does not quite solve the problem of constructing countably additive probability measures. It reduces it to constructing them on fields. The following theorem is important in the theory of Lebesgue integrals and is very useful for the construction of countably additive probability measures on the real line. The proof will again be only sketched. The natural σ-field on which to define a probability measure on the line is the Borel σ-field. This is defined as the smallest σ-field containing all intervals and includes in particular all open sets.

Let us consider the class of subsets of the real numbers, $I = \{I_{a,b}: -\infty \leq a < b \leq \infty\}$ where $I_{a,b} = \{x: a < x \leq b\}$ if $b < \infty$, and $I_{a,\infty} = \{x: a < x < \infty\}$. In other words I is the collection of intervals that are left-open and right-closed. The class of sets that are finite disjoint unions of members of I is a field F, if the empty set is added to the class. If we are given a function $F(x)$ on the real line which is nondecreasing, continuous from the right and satisfies

$$\lim_{x \to -\infty} F(x) = 0 \text{ and } \lim_{x \to \infty} F(x) = 1,$$

we can define a finitely additive probability measure P by first defining,

$$P(I_{a,b}) = F(b) - F(a)$$

for intervals and then extending it to F by defining it as the sum for disjoint unions from I. Let us note that the Borel σ-field B on the real line is the σ-field generated by F.

Lebesgue

Theorem: P is countably additive on F if and only if $F(x)$ is a right continuous function of x. Therefore for each right continuous nondecreasing function $F(x)$ with $F(-\infty) = 0$ and $F(\infty) = 1$ there is a unique probability measure P on the Borel subsets of the line, such that $F(x) = P(I_{-\infty,x})$. Conversely every countably additive probability measure P on the Borel subsets of the line comes from some F. The correspondence between P and F is one-to-one.

Proof: The only difficult part is to establish the countable additivity of P on F from the right continuity of $F(\cdot)$. Let $A_j \in F$ and $A_j \downarrow \Phi$, the empty set.

Let us assume, $P(A_j) \geq \delta > 0$, for all j and then establish a contradiction.

Step 1: We take a large interval $[-\ell, \ell]$ and replace A_j by Bj

$= A_j \cap [-\ell, \ell]$. Since $|P(A_j)-P(B_j)| \leq 1-F(\ell)+F(-\ell)$, we can make the choice of ℓ large enough that $P(B_j) \geq \frac{\delta}{2}$. In other words we can assumes without loss of generality that,

$$P(A_j) \geq \frac{\delta}{2}$$

and $A_j \subset [-\ell, \ell]$ for some fixed $\ell < \infty$.

Step 2: If,

$$A_j = \bigcup_{i=1}^{k_j} I_{a_{j,i}, b_{j,i}}$$

use the right continuity of F to replace A_j by B_j which is again a union of left open right closed intervals with the same right end points, but with left end points moved ever so slightly to the right. Achieve this in such a way that

$$P(A_j - B_j) \leq \frac{\delta}{10.2^j}$$

for all j.

Step 3. Define C_j to be the closure of B_j, obtained by adding to it the left end points of the intervals making up B_j. Let $E_j =$

$$\bigcap_{i=1}^{j} B_i \text{ and } D_j = \bigcap_{i=1}^{j} C_i .$$

Then,

- The sequence D_j of sets is decreasing,
- Each D_j is a closed bounded set,
- Since $A_j \supset D_j$ and $A_j \downarrow \Phi$, it follows that $D_j \downarrow \Phi$.

Because D_j É E_j and,

$$P(E_j) \geq \frac{\delta}{2} - \sum_j P(A_j - B_j) \geq \frac{4}{10\delta},$$

each D_j is nonempty and this violates the finite intersection property that every decreasing sequence of bounded nonempty closed sets on the real line has a nonempty intersection, *i.e.* has at least one common point. The rest of the proof is left as an exercise. The function F is called the distribution function corresponding to the probability measure P.

Example: Suppose $x_1, x_2 ..., x_n ...$ is a sequence of points and we have probabilities p_n at these points then for the discrete measure

$$P(A) = \sum_{n:x_n \in A} p_n$$

we have the distribution function,

$$F(x) = \sum_{n:x_n \leq x} p_n$$

that only increases by jumps, the jump at x_n being p_n. The points $\{x_n\}$ themselves can be discrete like integers or dense like the rationals.

Example: If $f(x)$ is a nonnegative integrable function with integral 1, *i.e.*,

$$\int_{-\infty}^{\infty} f(y)dy = 1 \text{ then } F(x) = \int_{-\infty}^{x} f(y)dy$$

is a distribution function which is continuous. In this case f is the density of the measure P and can be calculated as $f(x) = F'(x)$. There are (messy) examples of F that are continuous, but do not come from any density. More on this later.

Exercise: Let us try to construct the Lebesgue measure on the rationals $Q \subset [0, 1]$. We would like to have,

$$P[I_{a,b}] = b-a$$

for all rational $0 \leq a \leq b \leq 1$. Show that it is impossible by showing that $P[\{q\}] = 0$ for the set $\{q\}$ containing the single rational q while,

$$P[Q] = P\left[\bigcup_{q \in Q} \{q\}\right] = 1.$$

Where does the earlier proof break down? Once we have a countably additive probability measure P on a space (Ω, S), we will call the triple (Ω, S, P) a probabilty space.

INTEGRATION

An important notion is that of a random variable or a measurable function.

Definition: A random variable or measurable function is map $f: \Omega \to R$, *i.e.* a real valued function $f(\omega)$ on Ω such that for every Borel set $B \subset R$, $f^{-1}(B) = \{\omega : f(\omega) \in B\}$ is a measurable subset of Ω, *i.e* $f^{-1}(B) \in \Sigma$. A function that is measurable and satisfies $|f(\omega)| \leq M$ all $\omega \in \Omega$ for some finite M is called a bounded measurable function. The following statements are the essential steps in developing an integration theory. Details can be found in any book on real variables.

1. If $A \in \Sigma$, the indicator function A, defined as

$$1_A(\omega) = \begin{cases} 1 & if \omega \in A \\ 0 & if \omega \notin A \end{cases}$$

is bounded and measurable.

2. Sums, products, limits, compositions and reasonable elementary operations like min and max performed on measurable functions lead to measurable functions.

3. If $\{A_j: 1 \leq j \leq n\}$ is a finite disjoint partition of into measurable sets, the function,

$$f(\omega) = \sum_j c_j 1_{A_j}(\omega)$$

is a measurable function and is referred to as a 'simple' function.

4. Any bounded measurable function f is a uniform limit of simple functions. To see this, if f is bounded by M, divide $[-M, M]$ into n subintervals I_j of length $\frac{2M}{n}$ with midpoints c_j. Let

$$A_j = f^{-1}(I_j) = \{\omega: f(\omega) \in I_j\}$$

and,

$$f_n = \sum_{j=1}^{n} c_j 1_{A_j}.$$

Clearly f_n is simple, $\sup_\omega | f_n(\omega) - f(\omega) | - \frac{M}{n}$, and we are done.

5. For simple functions,

$$f = \sum c_j 1_{A_j}$$

the integral $\int f(\omega) dP$ is defined to be $\sum_j c_j P(A_j)$. It enjoys the following properties:

(a) If f and g are simple, so is any linear combination $af + bg$ for real constants a and b and

$$\int (af + bg) dP = a \int f dP + b \int g dP.$$

(b) If f is simple so is $|f|$ and,

$$\left|\int fdP\right| \leq \int |f| dP \leq \sup_{\omega} |f(\omega)|$$

6. If f_n is a sequence of simple functions converging to f uniformly, then ,

$$a_n = \int f_n dP$$

is a Cauchy sequence of real numbers and therefore has a limit a as $n \to \infty$. The integral $\int fdP$ of f is defined to be this limit a. One can verify that a depends only on f and not on the sequence f_n chosen to approximate f.

7. Now the integral is defined for all bounded measurable functions and enjoys the following properties.

(a) If f and g are bounded measurable functions and a, b are real constants then the linear combination $af + bg$ is again a bounded measurable function, and

$$\int (af + bg)dP = a\int fdP + b\int gdP.$$

(b) If f is a bounded measurable function so is $|f|$ and $\left|\int fdP\right| \leq \int |f| dP \leq \sup_{\omega} |f(\omega)|$.

(c) In fact a slightly stronger inequality is true. For any bounded measurable f,

$$\int |f| dP \leq P(\{\omega : |f(\omega)| > 0\}) \sup_{\omega} |f(\omega)|$$

(d) If f is a bounded measurable function and A is a measurable set one defines

$$\int_A f(\omega)dP = \int 1_A(\omega) f(\omega)dP$$

and we can write for any measurable set A,

$$\int fdP = \int_A fdP + \int_{A^c} fdP$$

In addition to uniform convergence there are other weaker notions of convergence.

Definition: A sequence f_n functions is said to converge to a function f everywhere or pointwise if

$$\lim_{n\to\infty} f_n(\omega) = f(\omega)$$

for every $\omega \in \Omega$.

In dealing with sequences of functions on a space that has a measure defined on it, often it does not matter if the sequence fails to converge on a set of points that is insignificant. For example if we are dealing with the Lebesgue measure on the interval [0, 1] and $f_n(x) = x_n$ then $f_n(x) \to 0$ for all x except $x = 1$. A single point, being an interval of length 0 should be insignificant for the Lebesgue measure.

Definition: A sequence f_n of measurable functions is said to converge to a measurable function f almost everywhere or almost surely (usually abbreviated as a.e.) if there exists a measurable set N with $P(N) = 0$ such that

$$\lim_{n\to\infty} f_n(\omega) = f(\omega)$$

for every $\omega \in N^c$.

Note that almost everywhere convergence is always relative to a probability measure.

Another notion of convergence is the following:

Definition: A sequence f_n of measurable functions is said to converge to a measurable function f 'in measure' or 'in probability' if

$$\lim_{n\to\infty} P[\omega : | f_n(\omega) - f(\omega) | \geq \varepsilon] = 0$$

for every $\varepsilon > 0$.

Let us examine these notions in the context of indicator functions of sets $f_n(\omega) = 1_{A_n}(\omega)$. As soon as $A \neq B$, $\sup_\omega | 1_A(\omega) - 1_B(\omega) | = 1$, so that uniform convergence never really takes place. On the other hand one can verify that $1_{A_n}(\omega) \to 1_A(\omega)$ for every ω if and only if the two sets

$$A_n = \bigcap_n \bigcup_{m\geq n} A_m$$

and

$$\liminf_n A_n = \bigcup_n \bigcap_{m\geq n} A_m$$

both coincide with A. Finally $1_{A_n}(\omega) \to 1_A(\omega)$ in measure if and only if,

$$\lim_{n\to\infty} P(A_n \Delta A) = 0$$

where for any two sets A and B the symmetric difference $A\Delta B$ is defined as,

$$A\Delta B = (A \cap B^c) \cup (A^c \cap B) = A \cup B \cap (A \cap B)^c.$$

It is the set of points that belong to either set but not to both. For instance $1_{A_n} \to \omega$ in measure if and only if $P(A_n) \to 0$.

Lemma: If $f_n \to f$ almost everywhere then $f_n \to f$ in measure.

Proof: $f_n \to f$ outside N is equivalent to

$$\bigcap_n \bigcup_{m\geq n} [\omega : | f_m(\omega) - f(\omega) | \geq \varepsilon] \subset N$$

for every $\varepsilon > 0$. In particular by countable additivity

$$P[\omega : | f_n(\omega) - f(\omega) | \geq \varepsilon] \leq P$$

$$\left[\bigcup_{m\geq n} [\omega : | f_m(\omega) - f(\omega) | \geq \varepsilon\right] \to 0$$

as $n\to\infty$ and we are done.

Bounded Convergence Theorem

Theorem: If the sequence $\{f_n\}$ of measurable functions is uniformly bounded and if $f_n \to f$ in measure as $n \to \infty$, then $\lim_{n\to\infty}$

$$\int f_n dP = \int f dP.$$

Proof: Since,

$$\left|\int f_n dP - \int f dP\right| = \left|\int (f_n - f) dP\right| \leq \int | f_n - f | dP$$

we need only prove that if $f_n \to 0$ in measure and $| f_n | \leq M$ then $\int | f_n | dP \to 0$. To see this,

$$\int | f_n | dP = \int_{|f_n|\leq\varepsilon} | f_n | dP +$$

$$\int_{|f_n|>\varepsilon} | f_n | dP \leq \varepsilon + MP[\omega : | f_n(\omega) > \varepsilon]$$

and taking limits,

$$\limsup_{n\to\infty} \int | f_n | dP \leq \varepsilon$$

and since $\varepsilon > 0$ is arbitrary we are done. The bounded convergence theorem is the essence of countable additivity. Let us look at the example of $f_n(x) = x^n$ on $0 \leq x \leq 1$ with Lebesgue measure. Clearly $f_n(x) \to 0$ a.e. and therefore in measure. While the convergence is not uniform, $0 \leq x^n \leq 1$ for all n and x and so the bounded convergence theorem applies.

In fact,

$$\int_0^1 x^n dx = \frac{1}{n+1} \to 0.$$

However if we replace x^n by nx^n, $f_n(x)$ still goes to 0 a.e., but the sequence is no longer uniformly bounded and the integral does not go to 0. We now proceed to define integrals of nonnegative measurable functions.

Definition: If f is a nonnegative measurable function we define,

$$\int fdP = \left\{\sup \int gdP : g \text{ bounded}, 0 \leq g \leq f\right\}.$$

Fatou's Lemma

Theorem: If for each $n \geq 1$, $f_n \geq 0$ is measurable and $f_n \to f$ in measure as $n \to \infty$ then

$$\int fdP \leq \liminf_{n\to\infty} \int f_n dP.$$

Proof: Suppose g is bounded and satisfies $0 \leq g \leq f$. Then the sequence $h_n = f_n \wedge g$ is uniformly bounded and

$$h_n \to h = f \wedge g = g.$$

Therefore, by the bounded convergence theorem,

$$\int gdP = \liminf_{n\to\infty} \int h_n dP$$

Since,

$$\int h_n dP \leq \int f_n dP$$

for every n it follows that

$$\int gdP \leq \liminf_{n\to\infty} \int f_n dP.$$

As g satisfying $0 \leq g \leq f$ is arbitrary and we are done.

Monotone Convergence Theorem

Corollary: If for a sequence $\{f_n\}$ of nonnegative functions, we have $f_n \uparrow f$ monotonically then

$$\int f_n dP \to \int f dP \text{ as } n \to \infty.$$

Proof: Obviously,

$$\int f_n dP \leq \int f dP$$

and the other half follows from Fatou's lemma.

Now we try to define integrals of arbitrary measurable functions. A nonnegative measurable function is said to be *integrable* if,

$$\int f dP < \infty.$$

A measurable function f is said to be integrable if $|f|$ is integrable and we define,

$$\int f dP = \int f^+ dP - \int f^- dP$$

where $f^+ = f \vee 0$ and $f^- = -f \wedge 0$ are the positive and negative parts of f. The integral has the following properties.

1. It is linear. If f and g are integrable so is $af + bg$ for any two real constants and,
$$\int (af + bg) dP = a \int f dP + b \int g dP.$$
2. $\left|\int f dP\right| \leq \int |f| dP$ for every integrable f.
3. If $f = 0$ except on a set N of measure 0, then f is integrable and,
$$\int f dP = 0.$$
In particular if $f = g$ almost everywhere then,
$$\int f dP = \int g dP.$$

Theorem: If $\phi(x)$ is a convex function of x and $f(\omega)$ and $\phi(f(\omega))$ are integrable then

$$\int \phi(f(\omega)) dP \geq \phi\left(\int f(\omega) dP\right)$$

Proof: We have seen the inequlity already for $\phi(x) = |x|$. The proof is quite simple. We note that any convex function can be represented as the supremum of a collection of affine linear functions.

$$\phi(x) = \sup_{(a,b)\in E} \{ax + b\}.$$

It is clear that if $(a, b) \in E$, then $a f(\omega) + b \leq (f(\omega))$ and on integration this yields $am + b \leq E[(f(\omega))]$ where $m = E[f(\omega)]$. Since this is true for every $(a, b) \in E$, in view of the represenattion, our theorem follows.

The Dominated Convergence Theorem

Theorem: If for some sequence $\{f_n\}$ of measurable functions we have $f_n \to f$ in measure and $| f_n(\omega) | \leq g(\omega)$ for all n and ω for some integrable function g, then,

$$\int f_n dP \to \int f dP \text{ as } n \to \infty.$$

Proof: $g + f_n$ and $g - f_n$ are nonnegative and converge in measure to $g + f$ and $g - f$ respectively. By Fatou's lemma

$$\liminf_{n\to\infty} \int (g + f_n) dP \leq \int (g + f) dP$$

Since,

$$\int g dP$$

is finite we can subtract it from both sides and get

$$\liminf_{n\to\infty} \int f_n dP \geq \int f dP.$$

Working the same way with $g - f_n$ yields,

$$\liminf_{n\to\infty} \int f_n dP \leq \int f dP$$

and we are done.

Exercise: Take the unit interval with the Lebesgue measure and define,

$$f_n(x) = n^{\alpha} 1_{\left[0, \frac{1}{n}\right]}(x).$$

Clearly $f_n(x) \to 0$ for $x \neq 0$. On the other hand,

$$\int f_n(x) dx = n^{\alpha - 1} \to 0$$

if and only if $\alpha < 1$. What is $g(x) = \sup_n f_n(x)$ and when is g integrable ?

If $h(\omega) = f(\omega) + ig(\omega)$ is a complex valued measurable function with real and imaginary parts $f(\omega)$ and $g(\omega)$ that are integrable we define,

$$\int h(\omega)dP = \int f(\omega)dP + i\int g(\omega)dP$$

TRANSFORMATIONS

A measurable space (Ω, B) is a set Ω together with a σ-field B of subsets of Ω.

Definition: Given two measurable spaces (Ω_1, B_1) and (Ω_2, B_2), a mapping or a transformation from $T: \Omega_1 \to 2$, *i.e.* a function $\omega_2 = T(\omega_1)$ that assigns for each point $\omega_1 \in \Omega_1$ a point $\omega_2 = T(\omega_1) \in \Omega_2$, is said to be measurable if for every measurable set $A \in B_2$, the inverse image

$$T^{-1}(A) = \{\omega_1 : T(\omega_1) \in A\} \in B_1.$$

Theorem: If $f: \Omega_2 \to R$ is a real valued measurable function on Ω_2, then $g(\omega_1) = f(T(\omega_1))$ is a measurable real valued function on (Ω_1, B_1). Moreover g is integrable with respect to P if and only if f is integrable with respect to Q, and

$$\int_{\Omega_2} f(\omega_2)dQ = \int_{\Omega_1} g(\omega_1)dP$$

Proof: If $f(\omega_2) = 1_A(\omega_2)$ is the indicator function of a set $A \in B_2$, the claim in equation is in fact the definition of measurability and the induced measure. We see, by linearity, that the claim extends easily from indicator functions to simple functions.

By uniform limits, the claim can now be extended to bounded measurable functions. Monotone limits then extend it to nonnegative functions. By considering the positive and negative parts separately we are done.

A measurable trnasformation is just a generalization of the concept of a random variable. We can either think of a random variable as special case of a measurable transformation where the target space is the real line or think of a measurable transformation as a random variable with values in an arbitrary target space.

The induced measure $Q = PT^{-1}$ is called the *distribution* of the random variable F under P. In particular, if T takes real values, Q is a probability distribution on R.

PRODUCT SPACES

Given two sets Ω_1 and Ω_2 the Cartesian product $\Omega = \Omega_1 \times \Omega_2$ is the set of pairs (ω_1, ω_2) with $\omega_1 \in \Omega_1$ and $\omega_2 \in \Omega_2$. If Ω_1 and Ω_2 come with σ-fields B_1 and B_2 respectively, we can define a natural σ-field B on Ω as the σ-field generated by sets (measurable rectangles) of the form $A_1 \times A_2$ with $A_1 \in B_1$ and $A_2 \in B_2$. This σ-field will be called the product σ-field.

Definition: The product -field B is the σ-field generaated by the field F.

Given two probability measures P_1 and P_2 on (Ω_1, B_1) and (Ω_2, B_2) respectively we try to define on the product space (Ω, B) a probability measure P by defining for a measurable rectangle $A = A_1 \times A_2$

$$P(A_1 \times A_2) = P_1(A_1) \times P_2(A_2)$$

and extending it to the field F of finite disjoint unions of measurable rectangles as the obvious sum.

Lemma: The measure P is countably additive on the field F.

Proof: For any set $E \in F$ let us define the section E_{ω_2} as

$$E_{\omega_2} = \{\omega_1 : (\omega_1, \omega_2) \in E\}.$$

Then $P_1(E_{\omega_2})$ is a measurable function of ω_2 (is in fact a simple function) and,

$$P(E) = \int_{\Omega_2} P_1(E_{\omega_2}) dP_2.$$

Now let $E_n \in F \downarrow \Phi$, the empty set. Then it is easy to verify that E_{n,ω_2} defined by,

$$E_{n,\omega_2} = \{\omega_1 : (\omega_1, \omega_2) \in E_n\}$$

satisfies $E_{n,\omega_2} \downarrow \Phi$ for each $\omega_2 \in \Omega_2$. From the countable additivity of P_1 we conclude that $P_1(E_{n,\omega_2}) \to 0$ for each $\omega_2 \in \Omega_2$ and since, $0 \leq P_1(E_{n,\omega_2}) \leq 1$ for $n \geq 1$, it follows from equation and the bounded convergence theorem that,

$$P(E_n) = \int_{\Omega_2} P_1(E_{n,\omega_2}) dP_2 \to 0$$

establishing the countable additivity of P on F.

By an application of the Caratheodory extension theorem we conclude that P extends uniquely as a countably additive measure to the σ-field B (product σ-field) generated by F. We will call this the Product Measure P.

Corollary: For any $A \in B$ if we denote by A_{ω_1} and A_{ω_2} the respective sections,

$$A_{\omega_1} = \{\omega_2 : (\omega_1, \omega_2) \in A\}$$

and,

$$A_{\omega_2} = \{\omega_1 : (\omega_1, \omega_2) \in A\}$$

then the functions $P_1(A_{\omega_2})$ and $P_2(A_{\omega_1})$ are measurable and

$$P(A) = \int P_1(A_{\omega_2})dP_2 = \int P_2(A_{\omega_1})\,dP_1.$$

In particular for a measurable set A, $P(A) = 0$ if and only if for almost all ω_1 with respect to P_1, the sections A_{ω_1} have measure 0 or equivalently for almost all ω_2 with respect to P_2, the sections A_{ω_2} have measure 0.

Proof: The assertion is clearly valid if A is rectangle of the form $A_1 \times A_2$ with $A_1 \in B_1$ and $A_2 \in B_2$. If $A \in F$, then it is a finite disjoint union of such rectangles and the assertion is extended to such a set by simple addition. Clearly, by the monotone convergence theorem, the class of sets for which the assertion is valid is a monotone class and since it contains the field F it also contains the σ-field B generated by the field F.

Warning. It is possible that a set A may not be *measurable* with respect to the product σ-field, but nevertheless the sections A_{ω_1} and A_{ω_2} are all measurable, $P_2(A_{\omega_1})$ and $P_1(A_{\omega_2})$ are measurable functions, but

$$(A_{\omega_2})dP_2 \neq \int P_2(A_{\omega_1})\,dP_1.$$

In fact there is a rather nasty example where $P_1(A_{\omega_2})$ is identically 1 whereas $P_2(A_{\omega_1})$ is identically 0. The next result concerns the equality of the double integral, (*i.e.* the integral with respect to the product measure) and the repeated integrals in any order.

Fubini's Theorem

Theorem: Let $f(\omega) = f(\omega_1, \omega_2)$ be a measurable function of ω on (Ω, B). Then f can be considered as a function of ω_2 for each fixed ω_1 or the other way around. The functions $g_{\omega_1}(\cdot)$ and $h_{\omega_2}(\cdot)$ defined respectively on Ω_2 and Ω_1 by,

$$g_{\omega_1}(\omega_2) = h_{\omega_2}(\omega_1) = f(\omega_1, \omega_2)$$

are measurable for each ω_1 and ω_2. If f is integrable then the functions $g_{\omega_1}(\omega_2)$ and $h_{\omega_2}(\omega_1)$ are integrable for almost all ω_1 and ω_2 respectively.

Their integrals,

$$G(\omega_1) = \int_{\Omega_2} g_{\omega_1}(\omega_2) dP_2$$

and

$$H(\omega_1) = \int_{\Omega_1} h_{\omega_2}(\omega_2) dP_1$$

are measurable, finite almost everywhere and integrable with repect to P_1 and P_2 respectively. Finally

$$\int f(\omega_1, \omega_2) dP = \int g(\omega_1) dP_1 = \int H(\omega_2) dP_2$$

Conversely for a nonnegative measurable function f if either G are H, which are always measurable, has a finite integral so does the other and f is integrable with its integral being equal to either of the repeated integrals, namely integrals of G and H.

Proof: The proof follows the standard pattern. It is a restatement of the earlier corollary if f is the indicator function of a measurable set A. By linearity it is true for simple functions and by passing to uniform limits, it is true for bounded measurable functions f. By monotone limits it is true for nonnegative functions and finally by taking the positive and negative parts seperately it is true for any arbitrary integrable function f.

Warning. The following could happen. f is a measurable function that takes both positive and negative values that is not integrable. Both the repeated integrals exist and are unequal. The example is not hard.

DISTRIBUTIONS AND EXPECTATIONS

Let us recall that a triplet (Ω, B, P) is a Probability Space, if Ω is a set, B is a σ-field of subsets of Ω and P is a (countably additive) probability measure on B. A random variable X is a real valued measurable function on (Ω, B). Given such a function X it induces a probability distribution on the Borel subsets of the line $\alpha = PX^{-1}$. The distribution function $F(x)$ corresponding to is obviously,

$$F(x) = ((-\infty, x]) = P[\omega: X(\omega) \leq x].$$

The measure α is called the distibution of X and $F(x)$ is called the distribution function of X. If g is a measurable function of the real variable x, then $Y(\omega) = g(X(\omega))$ is again a random variable and its distribution $\beta = PY^{-1}$ can be obtained as $\beta = \alpha g^{-1}$ from α. The *Expectation* or *mean* of a random variable is defined if it is integrable and

$$E[X] = E^P[X] = \int X(\omega)dP.$$

By the change of variables formula it can be obtained directly from α as,

$$E[X] = \int x d\alpha.$$

Here we are taking advantage of the fact that on the real line x is a very special real valued function. The value of the integral in this context is referred to as the expectation or mean of α. Of course it exists if and only if

$$\int |x| d\alpha < \infty$$

and,

$$\left| \int x d\alpha \right| \leq \int |x| d\alpha.$$

Similarly,

$$E(g(X)) = \int g(X(\omega))dP = \int g(x)d\alpha$$

and anything concerning X can be calculated from. The statement X is a random variable with distribution α has to be interpreted in the sense that somewhere in the background there is a Probability Space and a random variable X on it, which has for its distribution. Usually only α matters and the underlying (Ω, B, P) never emerges from the background and in a pinch we can always say Ω is the real line, B are the Borel sets, P is nothing but α and the random variable $X(x) = x$.

Some other related quantities are,

$$Var(X) = \sigma^2(X) = E[X^2] - [E[X]]^2.$$

$Var(X)$ is called the variance of X.

MEASURE THEORY PROBABILITY

The Riemann Integral the integral that is treated in freshman calculus applies to continuous functions. It can be extended a little beyond the class of continuous functions, but not very far. It can be used to define the lengths, areas, and volumes of sets in R, R^2, and R^3, provided those sets are reasonably nice, in particular not too irregularly shaped. In R^2, the Riemann Integral defines the area under the graph of a function by dividing the x-axis into a collection of small intervals.

On each of these small intervals, two rectangles are erected: one lies entirely inside the area under the graph of the function, while the other rectangle lies entirely outside the graph. The function is Riemann integrable (and its integral equals the area under its graph) if, by making the intervals sufficiently small, it is possible to make the sum of the areas of the outside rectangles arbitrarily close to the sum of the areas of the inside rectangles.

Measure theory provides a way to extend our notions of length, area, volume etc. to a much larger class of sets than can be treated using the Riemann Integral. It also provides a way to extend the Riemann Integral to Lebesgue integrable functions, a much larger class of functions than the continuous functions.

The fundamental conceptual difference between the Riemann and Lebesgue integrals is the way in which the partitioning is done. As noted above, the Riemann Integral partitions the domain of the function into small intervals.

By contrast, the Lebesgue Integral partitions the *range* of the function into small intervals, then considers the set of points in the domain on which the value of the function falls into one of these intervals. Let f: $[0, 1] \rightarrow R$.

Given an interval $[a, b) \subseteq R$, $f^{-1}([a, b))$ may be a very messy set. However, as long as we can assign a "length" or "measure" $\mu(f^{-1}([a, b)))$ to this set, we know that the contribution of this set to the integral of f should be between $a\mu(f^{-1}([a, b)))$ and $b\mu(f^{-1}([a, b]))$. By making the partition of the range finer and finer, we can determine the integral of the function.

Clearly, the key to extending the Lebesgue Integral to as wide a class of functions as possible is to define the notion of "measure" on as wide a class of sets as possible. In an ideal world, we would be able to define the measure of *every* set; if we could do this, we could then define the Lebesgue integral of *every* function. Unfortunately, as we shall see, it is not possible to define a measure with nice properties on every subset of R.

Measure theory is thus a *second best* exercise. We try to extend the notion of measure from our intuitive notions of length, area and volume to as large a class of *measurable* subsets of R, R^2, and R^3 as possible. In order to be able to make use of measures and integrals, we need to know that the class of measurable sets is closed under certain types of operations. If we can assign a sensible notion of measure to a set, we ought to be able to assign a sensible notion to its complement.

Probability and statistics focus on questions about convergence of sequences of random variables. In order to talk about convergence, we need to be able to assign measures to countable unions and countable intersections of measurable sets. Thus, we would like the collection of measurable sets to be a σ-algebra:

Definition: A *measure space* is a triple (Ω, B, μ), where

1. Ω is a set
2. B is a σ-algebra of subsets of Ω, *i.e.*
   (a) $B \subset 2^{\Omega}$, *i.e.* B is a collection of subsets of Ω
   (b) $\phi, \Omega \in B$
   (c) $B_n \in B,\ n \in N \Rightarrow \bigcup_{n \in N} \in B$
   (d) $B \in B \Rightarrow \backslash\ B \in B$
3. μ is a nonnegative, countably additive set function on B, *i.e.*
   (a) $\mu\colon B \to R^+ \cup \{\infty\}$
   (b) $B_n \in B,\ n \in N,\ B_n \cap B_m = \phi$ if $n \neq m \Rightarrow \mu$
   $$\left(\bigcup_{n \in N} B\right) = \sum_{n \in N} \mu(B_n) \text{ and } \mu(\phi) = 0.$$

Remark: The definition of a σ-algebra is closely related to the properties of open sets in a metric space. Recall that the collection of open sets is closed under (1) arbitrary unions and

(2) finite intersections; by contrast, a σ-algebra is closed under (1) countable unions and (2) countable intersections. Notice also that σ-algebras are closed under complements; the complement of an open set is closed, and generally not open, so closure under taking complements is not a property of the collection of open sets. The analogy between the properties of a σ-algebra and the properties of open sets in a metric space will be very useful in developing the Lebesgue integral. Recall that a function $f: X \to Y$ is continuous if and only if $f^{-1}(U)$ is open in X for every open set U in Y.

Recall from the earlier discussion that the Lebesgue integral of a function f is defined by partitioning the range of the fuction f into small intervals, and summing up numbers of the form $a\mu\,(f^{-1}([a, b)))$; thus, we will need to know that $f^{-1}([a, b)) \in B$. We will see in a while that a function $f: (\Omega, B, \mu) \to (\Omega', B')$ is said to be measurable if $f^{-1}(B) \in B$ for every $B' \in B'$. Thus, there is a close analogy between measurable functions and continuous functions.

As you know from calculus, continuous functions on a closed interval can be integrated using the so-called Riemann integral; the Lebesgue integral extends the Riemann integral to all bounded measurable functions (and many unbounded measurable functions).

Remark: Countable additivity implies $\mu(\phi) = 0$ provided there is some set B with $\mu(B) < \infty$; thus, the requirement $\mu(\phi) = 0$ is imposed to rule out the pathological case in which $\mu(B) = \infty$ for all $B \in B$.

Remark: If we have a finite collection $B_1 \ldots, B_k \in B$ with $B_n \cap B_m = \phi$ if $n \neq m$, we can write $B_n = \phi$ for $n > k$, and obtain,

$$\mu(B_1 \cup \ldots \cup B_k)$$

$$= \mu\left(\bigcup_{n\in N} B_n\right) = \sum_{n\in N} \mu(B_n)$$

$$= \sum_{n=1}^{k} \mu(B_k) + \sum_{n=k+1}^{\infty} \mu(\phi)$$

$$= \sum_{n=1}^{k} \mu(B_k),$$

so the measure is additive over finite collections of disjoint measurable sets.

Example: Suppose that we are given a partition $\{\Omega_\lambda : \lambda \in \Lambda\}$ of a set Ω, *i.e.*

$$\Omega = \bigcup_{\lambda \in \Lambda} \Omega_\lambda \text{ and } \lambda \neq \lambda' \Rightarrow \Omega_\lambda \bigcap \Omega_{\lambda'} \phi .$$

Then we can form a σ-algebra as follows: Let

$$B_\Lambda = \left\{ \bigcup_{\lambda \in C} \Omega_\lambda : C \subseteq \Lambda \right\}$$

In other words, B_Λ is the collection of all subsets of Ω which can be formed by taking unions of partition sets. B_Λ is closed under complements, as well as arbitrary (not just countable) unions and intersections. Suppose the partition is finite, *i.e.* Λ is finite, say $\Lambda = \{1..., n\}$. Then B_Λ is finite; it has exactly 2^n elements, each corresponding to a subset of Λ. Suppose now that Λ is countably infinite; since every subset $C \subseteq \Lambda$ determines a different set $B \in B_\Lambda$, B_Λ is uncountable. Suppose finally that Λ is uncountable. For concreteness, let $\Omega = R$, $\Lambda = R$, and $\Omega_\lambda = \{\lambda\}$, *i.e.* each set in the partition consists of a single real number. There are many $\tilde{A}$-algebras containing this partition. As before, we can take

$$B_\Lambda = \left\{ \bigcup_{\lambda \in C} \Omega_\lambda : C \subseteq \Lambda \right\} = \{C : C \subseteq R\} = 2^R$$

This is a perfectly good σ-algebra; however, as we indicated above (and as we shall show) it is not possible to define a measure with nice properties on this σ-algebra. The smallest σ-algebra containing the partition is

$$B0 = \{C : C \subseteq R,\ C \text{ countable or } R \setminus C \text{ countable}\}$$

This σ-algebra is too small to allow us to develop probability theory. We want a σ-algebra on R which lies between B_0 and B_Λ.

Definition: The Borel *σ-algebra* on R is the smallest σ-algebra containing all open sets in R. In other words, it is the σ-algebra

$$B = \cap \{C : C \text{ is a } \sigma\text{-algebra, } U \text{ open} \Rightarrow U \in C\}$$

i.e. it is the intersection of the class of all σ-algebras that contain all the open sets in R. A set is called *Borel* if it belongs to B.

Remark: Obviously, if U is an open set in R, then $U \in B$. If C is a closed set in R, then $R \setminus C$ is open, so $R \setminus C \in B$, so $C \in B$; thus, every closed set is a Borel set. Every countable intersection of open sets, and every countable union of closed sets, is Borel. Every countable set is Borel (exercise), and every set whose complement is countable is Borel. If $\{U_{nm}: n, m \in N\}$ is a collection of open sets, then,

$$\bigcup_n \left(\bigcap_m U_{nm}\right)$$

is a Borel set. Thus, Borel sets can be quite complicated.

The most important example of a measure space is the Lebesgue measure space, which comes in two flavors. The first flavour is (R, B, μ), where B is the Borel σ-algebra and μ is Lebesgue measure, the measure defined in the following theorem; the second flavour is (R, C, μ), where C is the σ-algebra defined in the proof sketch of the following theorem.

Theorem: If B is the Borel σ-algebra on R, there is a unique measure μ (called Lebesgue measure) defined on B such that $\mu((a, b)) = b-a$ provided that $b > a$. For every Borel set B,

$$\mu(B) = \sup\{\mu(K): K \text{ compact}, K \subseteq B\}$$
$$= \inf\{\mu(U): U \text{ open}, U \supseteq B\}$$

The proof works by gradually extending μ from the open intervals to the Borel σ-algebra. First, one shows one can extend μ to bounded open sets; it follows that one can extend it to compact sets. Then let $C_n = \{C \subset [-n, n]: \sup\{\mu(K): K \text{ compact}, K \subset C\} = \inf\{\mu(U): U \text{ open}, C \subset U\}\}$, and $C = \{C \subset R: C \cap [-n, n] \in C_n \text{ for all } n\}$. If $C \in C$, we define $\mu(C) = \sup\{\mu(K): K \text{ compact}, K \subset C\}$. One can verify that C is a σ-algebra containing every open set, hence $C \supset B$.

Definition: A measure space (Ω, B, μ) is *complete* if $B \mu B$, $A \subseteq B$, $\mu(B) = 0 \Rightarrow A \in B$. The *completion* of a measure space (Ω, B, μ) is the measure space $(\Omega, \overline{B}, \overline{\mu})$, where

$$\overline{B} = \{B \subseteq \Omega : \exists_{D,D \in B} C \subseteq B \subseteq D, \mu(D \setminus C) = 0\}$$

It is easy to verify that the Lebesgue measure space (R, C, μ) is complete, and is the completion of (R, B, μ), where B is the Borel σ-algebra.

Definition: Suppose μ is a measure on the Lebesgue σ-algebra C on R. μ is *translation-invariant* if, for all $x \in R$ and all $C \in C$, $\mu(C + x) = \mu(C)$.

Theorem: The Lebesgue measure space (R, C, μ) is translation-invariant.

The theorem follows readily from the construction of

Lebesgue measure, since translation doesn't change the length of intervals. Observe if $x \in R$, $\mu(\{x\}) \leq \mu((x-\varepsilon/2, x+\varepsilon/2)) = \varepsilon$ for every positive ε, so $\mu(\{x\}) = 0$.

As we have already indicated, it is not possible to extend Lebesgue measure to every subset of R, at least in the conventional set-theoretic foundations of mathematics.

Theorem: There is a set $D \subset R$ which is not Lebesgue measurable.

Proof: We actually prove the following stronger statement: there is no translation-invariant measure μ defined on all subsets of R such that $0 < \mu([0, 1]) < \infty$. Let μ be a translation-invariant measure defined on all subsets of R. Define an equivalence relation ~ on R by

$$x \sim y \Leftrightarrow x - y \in Q$$

Let D be formed by choosing exactly one element from each equivalence class, and such that all the chosen elements lie in $[0, 1)$.

Thus,

$$\forall_{x \in R} \exists_{d \in D} d - x \in Q$$
$$\forall_{d_1, d_2 \in D} d_1 - d_2 \notin Q$$

Given $x, y \in [0, 1)$, define,

$$x +' y = \begin{cases} x + y & \text{if } x + y \in [0,1) \\ x + y - 1 & \text{if } x + y \in [1,2) \end{cases}$$

The operation +' is *addition modulo one*. It is easy to check that, given any $C \subseteq [0, 1)$ and $y \in [0, 1)$,

$$\mu(C +' y) = \mu(C)$$

i.e. μ is translation-invariant with respect to translation using the operation +'. Then

$$[0, 1) = \bigcup_{q \in Q \cap [0,1)} D +' q$$

so,

$$\mu([0, 1)) = \sum_{q \in Q \cap [0,1)} \mu(D +' q) = \sum_{q \in Q \cap [0,1)} \mu(D)$$

Then either $\mu(D) = 0$, in which case $\mu([0, 1)) = 0$, so $\mu([0, 1]) = 0$, or $\mu(D) > 0$, in which case $\mu([0, 1)]) \geq \mu([0, 1)) = \infty$.

Definition: A measure space (Ω, B, μ) is a ***probability space*** if $\mu(\Omega) = 1$.

From now on, we will restrict attention to probability spaces.

Example: By abuse of notation, let C denote the collection of Lebesgue measurable sets which are subsets of [0, 1], and let μ be the restriction of

Lebesgue measure to this σ-algebra. ([0, 1], C, μ) is called the Lebesgue probability space.

Theorem: Let (Ω, B, μ) be a probability space. Suppose $B_1 \supseteq B_2 \supseteq \ldots \supseteq B_n \supseteq \ldots$ with $B_n \in B$ for all n.

Then,

$$\lim_{n\to\infty} \mu(B_n) = \mu\left(\bigcap_{n\in N} B_n\right)$$

In particular, if $\bigcap_{n\in N} B_n = \phi$, then $\lim_{n\to\infty} \mu(B_n) = 0$

Proof: Let $C_m = B_m \setminus B_{m+1}$. Then $C_n \cap C_m = \phi$ if $n \neq m$, and

$$B_1 = \left(\bigcap_{n\in N} B_n\right) \cup \left(\bigcup_{m\in N} C_m\right)$$

so,

$$\sum_{m\in N} \mu(C_m) = \mu(B_1) - \mu\left(\bigcap_{m\in N} B_n\right) \leq \mu(B_1) < \infty$$

so,

$$\sum_{m\in N} \mu(C_m) \text{ converges and hence}$$

$$\sum_{m=n}^{\infty} \mu(C_m) \to 0 \text{ as } n \to \infty$$

$$\mu(B_n) = \mu\left(\bigcap_{m\in N} B_m\right) + \sum_{m=n}^{\infty} \mu(C_m) \to \mu\left(\bigcap_{m\in N} B_m\right)$$

We now turn to the definition of a *random variable*. We think of Ω as the set of all possible states of the world tomorrow. Exactly one state will occur tomorrow. Tomorrow, we will be able to observe some function of the state which occurred. We do not know today which state will occur, and hence what value of the function will occur. However, we do know the probability that any given state will occur, and we know the mapping from possible states into values of the function, so we know the probability that the function will take on any given possible value. Thus, viewed from today, the value the function will take on tomorrow is a random variable; we know its probability distribution, but not its exact value. Tomorrow, we will observe the value which is realised.

Definition: Let (Ω, B, P) be a probability space. A random variable on Ω is a function $X: \Omega \to R \cup \{-\infty, \infty\}$ satisfying

$$P(\{\omega \in \Omega: X(\omega) \in \{-\infty, \infty\}\}) = 0$$

which is measurable, *i.e.* for every Borel set $B \subset R$, $X^{-1}(B) \in B$.

Observe the analogy beteen the definition of a measurable function and the characterization of a continuous function in terms of open sets: f is continuous if and only if $f^{-1}(U)$ is open for every open set U.

Lemma: Let (Ω, B, P) be a probability space. A function $X: \Omega \to R \cup \{-\infty, \infty\}$ satisfying

$$P(\{\omega \in \Omega: X(\omega) \in \{-\infty, \infty\}\}) = 0$$

is a random variable if and only if for every open interval (a, b), $X^{-1}((a, b)) \in B$.

Proof: Since every open interval is a Borel set, if X is a random variable, then,

$$X^{-1}((a, b)) \in B.$$

Now suppose that $X^{-1}((a, b)) \in B$ for every open interval (a, b). Consider,

$$M = \{B \subset R: X^{-1}(B) \in B\}$$

We claim that M is a σ-algebra. Note that $X^{-1}(\phi) = \phi \in B$, so, $\phi \in M$. $X^{-1}(R) = \Omega \setminus X^{-1}(\{-\infty, \infty\})$. Since $P(X^{-1}(\{-\infty, \infty\}))$ is zero (and hence defined), $X^{-1}(\{-\infty, \infty\}) \in B$; since B is a σ-algebra, it is closed under complements, so X–1 (R) Ω B, and RΩ M. If $B \in M$, $X^{-1}(B) \in B$, so $X^{-1}(R \setminus B) = X^{-1}(R) \setminus X^{-1}(B) \in B$ because $X^{-1}(R) \in B$ and $X^{-1}(B) \in B$. Therefore, $R \setminus B \in M$, so M is closed under complements.

Now suppose $B_n \in M$, $n \in N$. Then $X^{-1}(B_n) \in B$ for all n, so,

$$X^{-1}\left(\bigcup_{n \in N} B_n\right) = \bigcup_{n \in N} X^{-1}(B_n) \in B$$

and hence M is closed under countable unions. Therefore, M is a Ω-algebra. Since every open set is a countable union of open intervals, M contains every open set; since the Borel Ω-algebra is the smallest Ω-algebra containing the open sets in

R, M contains every Borel set, so X is measurable, and hence a random variable.

Theorem: If f: [0, 1] $\to R$ is continuous, then f is a random variable on the Lebesgue probability space ([0, 1], C, μ).

Proof: Since f takes values in R, $f^{-1}(\{-\infty, \infty\}) = \phi$. If (a, b) is an open interval, then since f is continuous, $f^{-1}((a, b))$ is an open subset of [0, 1], so $f^{-1}((a, b)) \in C$. By Lemma, f is a random variable.

Example: A random variable f: [0, 1] $\to R$ can be discontinuous everywhere. Consider $f(x) = 1$ if x is rational, 0 if x is irrational. f is discontinuous everywhere. Suppose B is a Borel set.

There are four cases to consider:

1. If $0, 1 \notin B$ $f^{-1}(B) = \phi \in C$.
2. If $0 \in B$, $1 \notin B$, $f^{-1}(B) = [0, 1] \setminus Q \in C$.
3. If $0 \notin B$, $1 \in B$, $f^{-1}(B) = [0, 1] \cap Q \in C$.
4. If $0, 1 \in B$, $f^{-1}(B) = [0, 1] \in C$.

Thus, f is measurable.

In elementary probability theory, random variables are not rigorously defined. Usually, they are described only by specifying their cumulative distribution functions. Continuous and discrete distributions often seem like entirely unconnected notions, and the formulation of mixed distributions (which have both continuous and discrete parts) can be problematic. Measure theory gives us a way to deal simultaneously with continuous, discrete, and mixed distibutions in a unified way.

We shall first define the cumulative distribution function of a random variable, and establish that it satisfies the defining properties of a cumulative distribution functions, as defined in elementary probability theory. We will then show (first in examples, then in a general theorem) that given any function F satisfying the defining properties of a cumulative distribution function, there is in fact a random variable defined on the Lebesgue probability space whose cumulative distribution function is F.

Definition: Given a random variable X: $(\Omega, B, P) \to R \cup \{-\infty, \infty\}$, the cumulative distribution function of X is the function F: R $\to$ [0, 1] defined by,

$$F(t) = P(\{\omega \in \Omega: X(\omega) \leq t\})$$

Theorem: If X is a random variable, its cumulative distribution function F satisfies the following properties:

1. $\lim_{t \to -\infty} F(t) = 0$ this is often abbreviated as F(−∞) = 0.
2. $\lim_{t \to -\infty} F(t) = 1$ this is often abbreviated as F(∞) = 1.
3. F is increasing, *i.e.* $s < t \Rightarrow F(t) \geq F(s)$
4. F is right-continuous, *i.e.* for all t, $\lim_{s \searrow t} F(s) = F(t)$

Proof: We prove only right-continuity, leaving the rest as an exercise. Since F is increasing, it is enough to show that (as an exercise, think through carefully why this *is* enough)

$$\lim_{n \to \infty} F\left(t + \frac{1}{n}\right) = F(t) \quad F\left(t + \frac{1}{n}\right) - F(t)$$

$$= P\left(\left\{\omega \in \Omega : X(\omega) \leq t + \frac{1}{n}\right\}\right) - P(\{\omega \in \Omega : X(\omega) \leq t\})$$

$$= P\left(X^{-1}\left(-\infty, t + \frac{1}{n}\right)\right) - P(\{X^{-1}((-\infty, t]))$$

$$= P\left(x^{-1}\left(t, t + \frac{1}{n}\right)\right) \to P\left(\bigcap_{n \in N} X^{-1}\left(t, t + \frac{1}{n}\right)\right)$$

$$= P\left(X^{-1}\left(\bigcap_{n \in N} t, t + \frac{1}{n}\right)\right)$$

$= P(X^{-1}(\phi))$

$= P(\phi) = 0$

Example: The *uniform distribution* on [0, 1] is the cumulative distribution function

$$F(t) = \begin{cases} 0 & \text{if} \quad t < 0 \\ t & \text{if} \quad t \in [0,1] \\ 1 & \text{if} \quad t > 1 \end{cases}$$

Consider the random variable X defined on the Lebesgue probability space X: $([0, 1], C, \mu) \to R$, $X(t) = t$. Observe that

$$\mu(\{\omega \in [0, 1] : X(\omega) \leq t\}) = \mu([0, t]) = t$$

Thus, X has the uniform distribution on [0, 1]. Notice also that F is strictly increasing and hence one-to-one on [0, 1]. Thus, $F|_{[0,1]}$ has an inverse function,

$$\left(F|_{[0,1]}\right)^{-1}:[0,1]\rightarrow[0,1] \text{ In fact, } X=\left(F|_{[0,1]}\right)^{-1}.$$

Example: The *standard normal distribution* has the cumulative distribution function

$$f(t)=\frac{1}{\sqrt{2\pi}}\int_{-\infty}^{t}e^{-x^2/2}dx$$

Notice that F is strictly increasing, hence one-to-one, and the range of F is (0, 1), so F has an inverse function X: $(0, 1) \rightarrow R$; extend X to [0, 1] by defining $X(0) = -\infty$ and $X(1) = \infty$. One can show that the inverse of a strictly increasing, continuous function is strictly increasing and continuous. Since $X|_{(0,1)}$ is continuous, it is measurable on the Lebesgue probability space. Observe that,

$$\mu(\{\omega \in [0, 1]: X(\omega) \leq t\}) = \mu X^{-1}((-\infty, t]) = F(t)$$

so X has the standard normal distribution.

Example: Consider the cumulative distribution function

$$F(t)=\begin{cases}0 & \text{if} \quad t<-1\\ \frac{1}{2} & \text{if} \quad t\in[-1,1]\\ 1 & \text{if} \quad t\geq 1\end{cases}$$

F is the cumulative distribution function of a random variable which takes on the values –1 and 1, each with probability $\frac{1}{2}$. We define a random variable on the Lebesgue probability space by

$$X(\omega)=\begin{cases}-\infty & \text{if} \quad \omega=0\\ -1 & \text{if} \quad \omega\in\left(0,\frac{1}{2}\right)\\ 1 & \text{if} \quad \omega\in\left(\frac{1}{2},1\right)\\ \infty & \text{if} \quad \omega=1\end{cases}$$

Clearly, the cumulative distribution function of X is F. Notice that F is weakly but not strictly increasing, so it is not one-to-one and hence does not have an inverse function. However, notice that X satisfies the following weakening of the definition of an inverse:

$$X(\omega) = \inf \{t\text{: } F(t) \geq \omega\}$$

Theorem: Let F: $R \rightarrow [0, 1]$ be an arbitrary function satisfying conclusions. There is a random variable X defined on the Lebesgue probability space whose cumulative distribution function is F.

Proof: Let

$$X(\omega) = \inf \{t\text{: } F(t) \geq \omega\}$$

Since, $\omega' > \omega \Rightarrow \{t\text{: } F(t) \geq \omega'\} \subseteq \{t\text{: } F(t) \geq \omega\} \Rightarrow \inf\{t\text{: } F(t) \geq \omega'\} \geq \inf \{t\text{: } F(t) \geq \omega\} \Rightarrow X(\omega')\ e''\ X(\omega)$.

X is increasing (not necessarily strictly). Since $\lim_{\sigma \to -\infty} F(s) = 0$ and $\lim_{t \to \infty} F(t) = 1$, if X^{-1} (0, 1), then there exist $s, t \in R$ such that $F(s) < \omega < F(t)$; since F is increasing, $-\infty < X < X(\omega) < \infty$.

Thus,

$$\mu(\{\omega\text{: } X(\omega) \in \{-\infty, \infty\}\}) \leq \mu\{0, 1\}) = 0$$

Since X is increasing, if (a, b) is an open interval in R, $X^{-1}((a, b))$ is an interval (not necessarily open) in [0, 1], and hence $X^{-1}((a, b)) \in C$. By Lemma, X is a random variable.

$$X(\omega) \leq t \Leftrightarrow \inf \{s\text{: } F(s) \geq \omega\} \leq t$$
$$\Leftrightarrow \exists_{s \leq t} F(s) \geq \omega$$
$$\Leftrightarrow F(t) \geq \omega$$

so,

$$\mu(\{\omega\text{: } X(\omega) \leq t\}) = \mu(\{\omega\text{: } \omega \leq F(t)\})$$
$$= \mu([0, F(t)])$$
$$= F(t)$$

In probability theory, the choice of a particular probability space is usually considered arbitrary and unimportant; you should choose a probability space which is convenient for the particular problem at hand, in particular one on which you can easily write down random variables with the desired joint distribution.

In what follows, we will allow (Ω, B, P) to be an arbitrary probability space, not necessarily the Lebesgue probability space. Next, we develop a notion of integration.

Definition: A simple function is a function of the form

$$f = \sum_{i=1}^{n} \alpha_i \chi_{B_i}$$

where each B_i is a measurable set in Ω and χ_{B_i} is the characteristic function of B_i:

$$\chi_{B_i}(\omega) = \begin{cases} 1 & \text{if} \quad \omega \in B_i \\ 0 & \text{if} \quad \omega \notin B_i \end{cases}$$

We define the integral of f by,

$$\int_{\Omega} f dP = \sum_{i=1}^{n} \alpha_i P(B_i)$$

A given simple function may be expressed in the form of Equation in more than one way; one can show that one gets the same value of the integral from all of these different expressions.

Given a nonnegative random variable f:

$$(\Omega, B, P) \to R_+ \cup \{\infty\},$$

we can construct a simple function which closely approximates it from below. Choose constants,

$$0 = \alpha_1 < \alpha_2 < .... < \alpha_n < \alpha_{n+1} = \infty$$

Let,

$$f_n = \sum_{i=1}^{n} \alpha_i \chi_{f^{-1}}([\alpha_i, \alpha_{i+1}))$$

Thus,

$$f_n(\omega) = \alpha_i \to f(\omega) \in [\alpha_i, \omega_{i+1})$$

Because f is a random variable, $f^{-1}([\alpha_i, \omega_{i+1})) \in B$, so f_n is a simple function.

Definition: If f:

$$(\Omega, B, P) \to R_+ \cup \{\infty\}$$

is a random variable, define,

$$\int_{\Omega} f dP = \sup\left\{\int_{\Omega} g dP : 0 \le g \le f, g \text{ simple}\right\}$$

The value of the integral may be ∞. We say that f is integrable if,

$$\int_{\Omega} f dP < \infty .$$

Example: Consider the function,

$$f(x) = \frac{1}{x}$$

on the Lebesgue probability space ([0, 1], *C*, μ). Let

$$f_n = \sum_{i=1}^{n} i\chi_{f^{-1}([i,i+1))} = \sum_{i=1}^{n} i\chi_{\left(\left(\frac{1}{i+1},\frac{1}{i}\right]\right)}$$

f_n is a simple function and $0 \leq f_n \leq f$.

$$\int f_n d\mu = \sum_{i=1}^{n} i\mu\left(\left(\frac{1}{i+1},\frac{1}{i}\right]\right)$$

$$= \sum_{i=1}^{n} i\left(\frac{1}{i} - \frac{1}{i+1}\right) = \sum_{i=1}^{n} \frac{1}{i+1} \rightarrow \infty \text{ as } n \rightarrow \infty$$

Thus,

$$\int_{[0,1]} f d\mu = \infty.$$

Definition: If $f: (\Omega, B, P) \rightarrow R \cup \{-\infty, \infty\}$ is a random variable, define $f_+(\omega) = \max\{f(\omega), 0\}$, $f_-(\omega) = -\min\{f(\omega), 0\}$. Note that $f = f_+ - f_-$ and $|f| = f_+ + f_-$. We say f is *integrable* if f_+ and f_- are both integrable, and we define

$$\int_{\Omega} f dP = \int_{\Omega} f_+ dP - \int_{\Omega} f_- dP$$

The following theorem, which we shall not prove, shows that the Lebesgue and Riemann integrals coincide when the Riemann integral is defined.

Theorem: Suppose that $f: [a, b] \rightarrow R$ is Riemann integrable (in particular, this is the case if f is continuous). Then f is Lebesgue integrable and,

$$\int_{[a,b]} f d\mu = \int_a^b f(t)dt$$

In other words, the Lebesgue integral of f equals the Riemann integral of f.

Definition: We say that a property holds *almost everywhere* (abbreviated a.e.) or *almost surely* (abbreviated a.s.) if the set of ω for which it does *not* hold is a set of measure zero.

Definition: Suppose f: $(\Omega, B, P) \to R$ is a random variable. If $B \in B$, we define,

$$\int_B f d\mu = \int_\Omega f \chi_B dP$$

where χ_B is the characteristic function of B.

Theorem: Suppose f, g: $(\Omega, B, P) \to R$ are integrable, $A, B \in B$, $A \cap B = \phi$, and $c \in R$.

Then,

1. $\int_B CfdP = c\int_B fdP$
2. $\int_B f + gdP = \int_B fdP + \int_B gdP$
3. $f \leq g$ a.e. on $B \Rightarrow fdP \leq \int_B gdP$
4. $\int_{A \cup B} fdP = \int_A fdP + \int_B fdP$

Definition: Suppose f, g: $(\Omega, B, P) \to R$ are random variables. Say $f \sim g$ if $f(\omega) = g(\omega)$ a.e. Let $[f]$ denote the equivalence class of f under the equivalence relation $\sim$, *i.e.* $[f] = \{g: (\Omega, B, P) \to R$: g is measurable, $g \sim f\}$. Define

$L^1(\Omega, B, P) = \{[f]: f: \Omega \to R, f$ a random variable, f integrable$\}$

$L^2(\Omega, B, P) = [f]: f: \Omega \to R, f$ a random variable, f^2 integrable$\}$

Theorem: If (Ω, B, P) is a probability space,

$$L^2(\Omega, B, P) \subseteq L^1(\Omega, B, P)$$

Definition: Given a metric space (X, d), the *completion* of (X, d) is a metric space $(\bar{X}, \bar{d})$ such that $(\bar{X}, \bar{d})$ is complete, $\bar{d}\,|_X = d$, and X is dense in $(\bar{X}, \bar{d})$. We are justified in calling $(\bar{X}, \bar{d})$ *the* completion of (X, d) because if $(\hat{X}, \hat{y})$ is another completion of (X, d), then there is an isomorphism ϕ:

$$(\bar{X}, \bar{d}) \to (\hat{X}, \hat{y})$$

such that $\phi(x) = x$ for every $x \in X$, and,

$$\hat{d}\,(\phi(\bar{x}), \phi(\hat{y})) = \bar{d}(\bar{x}, \bar{y}) \text{ every } \bar{x}, \bar{y} \in \bar{X}.$$

Theorem: $L^1(\Omega, B, P)$ and $L^2(\Omega, B, P)$ are Banach spaces under the respective norms,

$$\| f \|_1 = \int_\Omega | f(\omega) | dP$$

$$\| f \|_2 = \sqrt{\int_\Omega f^2(\omega) dP}$$

For the Lebesgue probability space, $L^1([0, 1], C, \mu)$ and $L^2([0, 1], C, \mu)$ are the completions of the normed space C([0, 1]), with respect to the norms $\| . \|_1$ and $\| \cdot \|_2$.

Example: Recall that C([0, 1]) is a Banach space with the sup norm $f = \sup\{| f(x) | : x \in [0, 1]\}$. $(C[0, 1], \| \cdot \|_1)$ and $(C[0, 1], \| \cdot \|_2)$ are normed spaces, but they are not complete in these norms, hence are not Banach spaces. To see this, let $f_n(x)$ be the function which is zero for,

$$x \in \left[0, \frac{1}{2} - \frac{1}{n}\right], \text{ one for } x \in \left[\frac{1}{2} + \frac{1}{n}, 1\right]$$

and linear for,

$$x \in \left[\frac{1}{2} - \frac{1}{n}, \frac{1}{2} + \frac{1}{n}\right].$$

f_n is not Cauchy with respect to $\| \cdot \|_\infty$. However, f_n *is* Cauchy and has a natural limit with respect to $\| \cdot \|_1$. Indeed, let

$$f(x) = \begin{cases} 0 & \text{if} \quad x < \dfrac{1}{2} \\ 1 & \text{if} \quad x \geq \dfrac{1}{2} \end{cases}$$

Then,

$$\int_{[0,1]} | f_n - f | d\mu \leq \frac{1}{2} \times \frac{2}{n} = \frac{1}{n} \to 0$$

so f_n converges to f in $L^1([0, 1], C, \mu)$, and hence must be Cauchy with respect to $\| \cdot \|_1$. Notice that this limit does not belong to C([0, 1]).

Definition: For $f \in L^1(\Omega, B, P)$, define the *mean* or *expectation* of f by

$$E(f) = \int_\Omega f dP$$

For $f \in L^2(\Omega, B, P)$, define the variance of f by,

$$\text{Var}(f) \int_\Omega (f(\omega) - E(f)^2 dP$$

Definition: If $f, g \in L^2 (\Omega, B, P)$, define the *inner product* of f and g by,

$$f \cdot g = \int_\Omega f(\omega) g(\omega) dP$$

The properties of the inner product are closely analogous to those of the dot product of vectors in Euclidean space R^n. In particular, they determine a geometry, including lengths,

$$\left(\| f \|_2 = \sqrt{f \cdot f} \right)$$

and angles. The most basic property of the dot product, the Cauchy-Schwarz inequality, extends to the inner product.

Cauchy-Schwarz Inequality

Theorem: If $f, g \in L^2(\Omega, B, P)$,

$$| f \cdot g | \leq \| f \|_2 \| g \|_2$$

Definition: If $f, g \in L^2(\Omega, B, P)$, define the *covariance* of f and g by,

$$\text{Covar } (f, g) = E((f - E(f))(g - E(g)))$$

Proposition: If $f, g – L^2(\Omega, B, P)$, then

$$\text{Covar}(f, g) = E(fg) - E(f)\, E(g)$$

Observe from the definition that Covar (f, g) = Covar (g, f). Thus, given $f_1 \ldots, f_n – L^2(\Omega, B, P)$, the covariance matrix C whose (i, j) entry is c_{ij} = Covar(f_i, f_j) is a symmetric matrix. Hence, there is an orthonourmal basis of R^n composed of eigenvectors of C; expressed in this basis, the matrix becomes diagonal.

Observe also that,

$$\text{Covar}\left(\sum_{i=1}^{n} \alpha_i f_i, \sum_{j=1}^{n} \beta_j f_j, \right) = (\alpha_1, \ldots, \alpha_n) C \begin{pmatrix} \beta_1 \\ \vdots \\ \beta_n \end{pmatrix}$$

is a quadratic form of the type we studied earlier. One very useful consequence of the inner product structure on L^2 is the existence of orthogonal projections. Given a linearly independent family

$$f_1 \ldots, f_n – L^2(\Omega, B, P),$$

let V be the vector space spanned by $f_1 \ldots, f_n$. Then any $g \in L^2$

can be written in a unique way as,

$$g = \pi(g) + \omega,$$

where

$$\pi(v) = \sum_{i=1}^{n} \alpha_i f_i \in V$$

and $\omega \cdot v = 0$ for all $v \in V$ (in particular, $\omega \cdot f_i = 0$ for each i). $\pi(g)$ is also characterized as the point in V closest to g. The coefficients $\alpha_1 \ldots, \alpha_n$ are the regression coefficients of g on $f_1 \ldots, f_n$. If $f_1 \ldots, f_n$ are orthonourmal (*i.e.* $f_i \cdot f_j = 1$ if $i = j$ and 0 if $i \neq j$), $\alpha_i = g \cdot f_i$ for each i.

One frequently encounters two measures living on a single probability space. For example, Lebesgue measure is one measure on the real line R; any random variable determines another measure on R by its distribution.

Definition: Let μ and v be two measures on the same measurable space (X, A), *i.e.* (X, A, μ) and (X, A, v) are measure spaces.

We say v is *absolutely continuous with respect to* μ if,

$$B \in A,\ \mu(B) = 0 \Rightarrow v(B) = 0$$

We say μ is σ-finite if there exist $B_1, B_2 \ldots \in A$ such that,

$$X = \bigcup_{n \in N} B_n \text{ and } \mu(B_n) < \infty \text{ for all } n.$$

Radon-Nikodym

Theorem: Suppose that μ and v are two σ-finite measures on the same measurable space (X, A). Then v is absolutely continuous with respect to μ if and only if there exists $f \in L^1(X, A, \mu)$, $f \geq 0$ almost everywhere, such that,

$$v(B) = \int_B f d\mu$$

for all $B \in A$.

The Radon-Nikodym theorem tells us that if a random variable Z has a continuous distribution (in other words, the measure v determined by its cumulative distribution function F is absolutely continuous with respect to Lebesgues measure), then Z has a density function f, namely f is the Radon-Nikodym derivative of v with respect to Lebesgue measure.

We now turn to products of probability spaces (X, A, P) and (Y, B, Q). If $A \in A$ and $B \in B$, it is natural to define,

$$(P \times Q)(A \times B) = P(A)Q(B)$$

However, most of the subsets of $X \times Y$ we are interested in are not of the form $A \times B$ with $A \in A$ and $B \in B$. For example, if $X = Y = [0, 1]$, the diagonal $\{(x, y): x = y\}$ is *not* of that form. However, it is possible to write the diagonal in the form,

$$\bigcap_{m \in N} \bigcup_{n=1}^{i_m} A_{mn} \times B_{mn}$$

with $A_{mn} \in A$ and $B_{mn} \in B$. This suggests we should extend $P \times Q$ to the smallest σ-algebra containing $\{A \times B: A \in A, B \in B\}$.

Definition: Suppose (X, A, P) and (Y, B, Q) are probability spaces. The product probability space is $(X \times Y, A \times B, P \times Q)$, where

1. $A \times' B$ is the smallest σ-algebra containing all sets of the form $A \times B$, where $A \in A$ and $B \in B$
2. $P \times' Q$ is the unique measure on $A \times' B$ that takes the values $P(A)Q(B)$ on sets of the form $A \times B$, where $A \in A$ and $B \in B$.
3. $(X \times Y, A \times B, P \times Q)$ is the completion of $(X \times Y, A \times' B, P \times' Q)$

The existence of the product measure $P \times Q$ is proven by extending the definition of $P \times Q$ for sets of the form $A \times B$ to $A \times B$, in a manner similar to the process by which Lebesgue measure is obtained from lengths of intervals.

Fubini

Theorem: Suppose (X, A, P) and (Y, B, Q) are complete probability spaces. If,

$$f: X \times Y \rightarrow R \text{ is } A \times B$$

measurable and integrable, then,

1. For almost all $x \in X$, the function defined by
$$f_x(y) = f(x, y) \text{ is integrable in } Y;$$
2. For almost all $y \in Y$, the function defined by
$$f_y(x) = f(x, y) \text{ is integrable in } X;$$
3. $\int_Y f(x, y)dQ(y)$ is an integrable function of $x \in X$;
4. $\int_X f(x, y)dP(x)$ is an integrable function of $y \in Y$;

5. $\int_{X \times Y} f(x, y) d(P \times Q)(x, y)$

There are many notions of convergence of functions, and results showing that the integral of a limit of functions is the limit of the integrals.

Definition: Suppose,

$f_n, f: (\Omega, B, P) \to R$ are random variables.

1. We say f_n converges to f in probability or in measure if

$$\forall_{\varepsilon>0} \exists_N n > N \Rightarrow P(\{\omega : | f_n(\omega) | > \varepsilon\}) < \varepsilon$$

2. We say f_n converges to f almost everywhere (abbreviated a.e.) or *almost surely* (abbreviated *a.s.*) if

$$P(\{\omega: f_n(\omega) \to f(\omega)\}) = 1$$

Theorem: If f_n converges to f almost surely, then f_n converges to f in probability.

In the lecture, I gave an example of a sequence of functions which converges in probability, but not almost surely. The example is much easier to describe with pictures than with words, so I won't give the details here.

The idea is to construct a sequence of functions,

$$f_n: [0, 1] \to R$$

so that,

$$\mu(\{x: f_n(x) \neq 0\}) \to 0;$$

hence f_n converges in probability to the identically zero function f. However, we can move the set on which f_n is not zero around so that, for every $x \in [0, 1]$, $f_n(x) = 1$ for infinitely many n, so almost surely, $f_n(x)$ fails to converge to $f(x)$.

Example: Although convergence almost surely is a strong notion of convergence of functions, it is not sufficient for convergence of the integrals of the functions to the integral of the limit.

For example, consider f_n: $[0, 1] \to R$ defined by $f_n = n\chi_{[0,1/n]}$, which takes the value n on $[0, 1/n]$ and 0 everywhere else. f_n converges almost surely to f, the function which is identically zero, but if P is Lebesgue measure,

$$[0\ 1] f_n\, dP = n \times \frac{1}{n} = 1, \text{ while} \int_{[0,1]} f dP = 0.$$

Roughly speaking, f_n is converging to a function which is ∞ with probabilty 0, and in the Lebesgue integral, $0 \times \infty = 0$; some of the mass in f_n is lost in the limit operation. The next result tells us that, for nonnegative functions, this is the only thing that can go wrong, and the limit of the integrals is always at least as big as the integral of the limit function.

Bibliography

Association, A.: *Journal of the American Statistical Association*, London: Academic Press, 2002.

Company, R.: *Pure and Applied Science*, New York: Columbia University Press, 2009.

Hardle, W.: *Statistical Methods for Biostatistics and Related Fields*, New York: Columbia University Press, 2007.

Levy, B.: *Library of Congress Catalog*, CA: Sage Publications, 2008.

Mahajan, T.: *Methods in Biostatistics*, MA: Harvard University Press, 2006.

Mitcham, A.: *Encyclopedia of Science, Technology, and Ethics*, N.J.: Scarecrow Press, 2005.

Mitcham, C.: *Occupational and Environmental Health*, Boston: Allyn and Bacon, 2003.

Plichta, S.: *Statistics for Nursing and Allied Health*, Carbondale: Southern Illinois University Press, 2009.

Rao, G.: *Rational Medical decision Making: a Case-Based Approach*, Beverly Hills: Sage Publications, 2002.

Redmond, C.: *Biostatistics in Clinical Trials*, Albany: State University of New York Press, 2008.

Rimm, A.: *Basic Biostatistics in Medicine and Epidemiology*, New York: Oxford University Press, 2004.

Ropella, K.: *Introduction to Statistics for Biomedical Engineers*, NJ: L. Erlbaum Publications, 1998.

Samuel , P.: *Biostatistics: principles and practice*, Washington: CQ Press, 2004.

Schefler , W.: *Statistics for the Biological Sciences,* Philadelphia: Open University Press, 2003.

Schmitz, P.: *Internal Medicine: Just the Facts,* Ann Arbor: University of Michigan Press, 2001.

Slome, C.: Basic Epidemiological Methods and Biostatistics, N.J.: Erlbaum Publications, 1999.

Troidl, H.: *Surgical Research: Basic Principles and Clinical Practice,* Buckingham: Open University Press, 2000.

Varavikova, E.: *The new Public Health,* New York: Wiley Publications, 2000.

Walden, G.: *Polling and Survey Research Methods,* NJ: Princeton University Press, 2006.

Wilson , H.: *Bulletin of the Public Affairs Information Service,* New York: Holt, Rinehart, and Winston, 2005.

Index

A

Abbreviation 2, 3
Aboriginal 222
Accordance 37, 39, 97, 214
Acquaintance 112
Adequate 2, 56, 128, 238, 246
Aforementioned 157
Apparent 4, 30, 40, 41, 81, 161, 188
Approximated 15, 108, 250

B

Bimodal 121, 126, 184, 185, 186
Binomial 48, 57, 58, 59, 92, 93, 199, 207, 208, 209, 210, 211, 212, 215
Boxplot 145, 146, 186

C

Combined 20, 109, 132, 151, 152, 153, 154, 167, 168
Concordance 30, 49, 50, 52, 55, 65
Culminating 145

D

Demonstrated 18, 47, 53, 56, 210, 213
Deviate 54, 69, 70, 71, 77, 79, 80, 96, 97, 102, 103, 104, 107, 108, 111
Deviation 33, 51, 52, 54, 60, 62, 63, 65, 69, 71, 72, 74, 76, 77, 78, 79, 80, 82, 83, 96, 98, 99, 101, 102, 104, 106, 108, 109, 123, 129, 130, 131, 132, 133, 134, 135, 136, 137, 138, 139, 140, 142, 143, 144, 145, 146, 150, 151, 152, 153, 156, 175, 183, 211
Diagnostic 194
Dichotomous 85, 92
Discordance 47, 49, 50, 51, 52, 54, 57, 61
Discrepancies 48, 49, 50, 51, 58, 59, 60, 61, 62, 64, 194
Discrete 91, 258, 259, 280
Distortion 31, 249
Divvy 179

E

Endpoints 183, 184
Equivalent 31, 43, 52, 89, 109, 110, 111, 263
Established 37, 55, 58, 68, 94, 188, 211, 215, 220, 221, 226
Exponents 1, 2, 6, 103

G

Gained 79

Glance 10, 57, 71, 127
Glorious 157
Grasp 32, 53, 114, 159

H

Histogram 34, 37, 61, 63, 67, 81, 95, 102, 110, 183, 184, 185, 186, 187

I

Impartial 31
Integral 4, 6, 53, 55, 93, 107, 199, 200, 259, 260, 261, 264, 265, 269, 270, 271, 272, 273, 274, 284, 285, 291, 292

L

Laplace 103, 188, 200, 208

M

Mannheim 10, 13
Mantissa 7, 9, 10

O

Ordinal 84, 85, 86, 87, 89

P

Partialing 157, 158, 159, 166, 172, 178, 179
Percentile 85, 121, 130, 132, 145
Phenomenon 195, 196, 197, 210
Precisely 37, 40, 78, 98, 107, 142, 145, 254
Prominence 103

Q

Quetelet 103

R

Randomly 36, 48, 61, 92, 122, 124, 126, 147, 149, 152, 159, 160, 164, 198, 225, 226, 227, 230, 234, 235, 240, 242, 243, 244, 251
Rapid 36

S

Seriation 34, 35, 36, 37, 38, 39, 40, 45, 46, 67, 72, 95
Sigma 31, 115
Skewed 119, 121, 126, 127, 128, 137, 141, 172, 178, 185, 187
Sophomores 86
Squares 20, 21, 71, 77, 78, 79, 139, 140, 150, 151, 152, 153, 154, 167, 168, 170
Stratification 233, 234

T

Tendency 33, 95, 113, 114, 119, 122, 124, 126Tenured 124
Theoretical 47, 48, 49, 50, 52, 55, 56, 57, 58, 59, 60, 61, 63, 97, 101, 146, 190, 191, 193, 194

V

Variate 30, 32, 33, 35, 36, 38, 39, 40, 46, 68, 69, 70, 73, 78, 79, 91, 93, 104